Colloids in Paints

Tharwat F. Tadros

Colloids and Interface Science Series
edited by Tharwat F. Tadros

Other Volumes

Volume 1: Colloid Stability Part I

ISBN 978-3-527-31462-1

Volume 2: Colloid Stability Part II

ISBN 978-3-527-31503-1

Volume 3: Colloid Stability and Application in Pharmacy

ISBN 978-3-527-31463-8

Volume 4: Colloids in Cosmetics and Personal Care

ISBN 978-3-527-31464-5

Volume 5: Colloids in Agrochemicals

ISBN 978-3-527-31465-2

Complete Set

Colloids and Interface Science Series

Volumes 1–6

ISBN 978-3-527-31461-4

Colloids in Paints

Colloids and Interface Science Vol. 6

Tharwat F. Tadros

WILEY-VCH

WILEY-VCH Verlag GmbH & Co. KGaA

The Author

Prof. Dr. Tharwat F. Tadros
89 Nash Grove Lane
Wokingham, Berkshire RG40 4HE
United Kingdom

Library of Congress Card No.: applied for

British Library Cataloguing-in-Publication Data
A catalogue record for this book is available from the British Library.

Bibliographic information published by the Deutsche Nationalbibliothek
The Deutsche Nationalbibliothek lists this publication in the Deutsche Nationalbibliografie; detailed bibliographic data are available on the Internet at http://dnb.d-nb.de.

Printed in the Federal Republic of Germany
Printed on acid-free paper

Cover Design Adam-Design, Weinheim
Typesetting Asco Typesetters, Hong Kong
Printing and Binding betz-druck GmbH, Darmstadt

ISBN: 978-3-527-31466-9

Dedication

This book is dedicated to our grandchildren: Nadia, Dominic, Theodore, Bruno and Viola.

Contents

Colloids and Interface Science Series, Vol. 6
Colloids in Paints. Tharwat F. Tadros

ISBN: 978-3-527-31466-9

Preface

Paints or surface coatings are complex, multi-phase, colloidal systems that are applied as a continuous layer to a surface. The main purpose of a paint or surface coating is to provide aesthetic appeal, in addition to protecting the surface. When considering a paint formulation, one must know the specific interaction forces between the paint components and substrates. The main objectives of the present volume is to consider the fundamental colloid and interface science principles that can be applied in the formulation of a paint system and its application to various substrates. The first chapter gives a general introduction to the properties of the main components in a paint formulation, namely the disperse particles and the medium in which they are dispersed (the film formers and the solvent). A brief description of the properties of the dispersed particles and the medium (the continuous phase) is given. The second chapter deals with emulsion polymers (latexes), which are the most commonly used film formers in the coating industry. This is particularly the case with aqueous emulsion paints that are used for home decoration. These aqueous emulsion paints are applied at room temperature and the latexes coalesce on the substrate, forming a thermoplastic film. Sometimes functional polymers are used for cross-linking in the coating system. The polymer particles are typically submicron (0.1–0.5 μm). Chapter 3 deals with pigment dispersion. The state of dispersion of a pigment in a paint is vital as it determines its optical properties (e.g. color), flow properties (rheology), durability, opacity, gloss and storage stability. The role of surfactants (or polymers) in the dispersion process is considered when dealing with wetting of the powder by the liquid, breaking of the aggregates and agglomerates and comminution (milling) of the resulting particles into smaller units. Chapter 4 describes the colloid stabilization of the disperse phase in any paint formulation, which is essential during preparation of the pigment dispersion in order to prevent any particle reaggregation. The colloid stability is very important for the long-term storage of a paint formulation. Aggregation of particles leads to poor performance of the paint, such as its opacity, color and durability. Both electrostatic and steric stabilization are described at a fundamental level. A section on the adsorption and conformation of polymeric surfactants at the solid/liquid interface is given. This is key to understanding how these polymeric surfactants can stabilize the particles against aggregation. Finally, the interaction between particles containing

Colloids and Interface Science Series, Vol. 6
Colloids in Paints. Tharwat F. Tadros

ISBN: 978-3-527-31466-9

an adsorbed layer is discussed and the factors responsible for steric stabilization are considered. Chapter 5 describes the deposition of particles on substrates and their adhesion to them. The deposition of particles on surfaces is an important process in paint application, since it governs the paint film properties. Two main types of particles must be considered, namely the latex particles (which may be considered as emulsions) that determine film formation and the pigment particles such as TiO_2 and other colored pigments. Particle deposition is followed by their adhesion and it is essential to have uniform deposition (without any flocculation) to ensure that the final adhered film has the required properties, such as gloss and durability. Chapter 6 describes the basic principles of rheology. One of the most important characteristics of any paint formulation is its flow properties (rheology), which are very important for both its long-term physical stability and its application. In the latter case, it is important to control the rheology for ease of application and production of the required paint film. Two types of rheological techniques are described, namely steady-state and viscoelastic methods. Chapter 7 presents the fundamental principles of the rheology of suspensions, emulsions and their mixtures (suspoemulsions), which are the main components of any paint formulation. Chapter 8 describes rheology modifiers, thickeners and gels. In any paint formulation, one needs to modify the rheology of the system to achieve the following properties: (i) long-term physical stability (absence of creaming or sedimentation and separation) and (ii) ease of application, good coating and lack of sag. Chapter 9 describes the rheology of the complex paint formulation. Control of paint rheology is essential for successful utilization of the paint. Whatever application technique is used, such as spray gun, brush or roller, three stages must be taken into account when considering the rheology of a paint: (i) transfer of the paint from the bulk container to the applicator, (ii) transfer of the paint to the surface to form a thin and even film, namely film formation, and (iii) flow-out of the film surface, coalescence of the polymer latex particles and loss of the medium by evaporation. Each of these processes requires accurate control of the rheological characteristics. The latter are described at a fundamental level.

This volume on *Colloids in Paints* should give the formulator the basic fundamental principles that are involved in the various interactions that occur in such complex systems. The book should enable the physical chemist to formulate the paint using a rational approach rather than by 'trial and error'. It should also enable the formulator to apply the relevant techniques for accurate control of the paint system. The book is also intended for researchers involved in designing a paint system.

December 2009 *Tharwat Tadros*

1
Colloids in Paints

Paints or surface coatings are complex, multi-phase, colloidal systems that are applied as a continuous layer to a surface [1]. A paint usually contains pigmented materials to distinguish it from clear films that are described as lacquers or varnishes. The main purpose of a paint or surface coating is to provide aesthetic appeal and to protect the surface. For example, a motor car paint can enhance the appearance of the car body by providing color and gloss and it also protects the car body from corrosion.

When considering a paint formulation, one must know the specific interaction between the paint components and substrates. This subject is of particular importance when one considers the deposition of the components on the substrate and their adhesion to it. The substrate can be wood, plastic, metal, glass, etc. The interaction forces between the paint components and the substrate must be considered when formulating any paint. In addition, the method of application can vary from one substrate and another.

For many applications, it has been recognized that achieving a required property, such as durability, strong adhesion to the substrate, opacity, color, gloss, mechanical properties, chemical resistance and corrosion protection, requires the application of more than one coat. The first two or three coats (referred to as the primer and undercoat) are applied to seal the substrate and provide strong adhesion to the substrate. The topcoat provides the aesthetic appeal such as gloss, color and smoothness. This clearly explains the complexity of paint systems, which require a fundamental understanding of the processes involved such as particle–surface adhesion, colloidal interaction between the various components and mechanical strength of each coating.

The main objective of the present text is to consider the colloidal phenomena involved in a paint system, its flow characteristics or rheology, its interaction with the substrate and the main criteria that are needed to produce a good paint for a particular application.

To obtain a fundamental understanding of the above basic concepts, one must first consider the paint components. Most paint formulations consist of disperse systems (solid in liquid dispersions). The disperse phase consists of primary pigment particles (organic or inorganic), which provide the opacity, color and other optical effects. These are usually in the submicron range. Other coarse particles

Colloids and Interface Science Series, Vol. 6
Colloids in Paints. Tharwat F. Tadros

ISBN: 978-3-527-31466-9

(mostly inorganic) are used in the primer and undercoat to seal the substrate and enhance adhesion of the topcoat. The continuous phase consists of a solution of polymer or resin which provides the basis of a continuous film that seals the surface and protects it from the outside environment. Most modern paints contain latexes, which are used as film formers. These latexes (with a glass transition temperature mostly below ambient temperature) coalesce on the surface and form a strong and durable film. Other components may be present in the paint formulation such as corrosion inhibitors, driers and fungicides.

This introductory chapter gives a brief account of the properties of the main components in a paint formulation, namely the disperse particles and the medium in which they are dispersed (the film formers and the solvent).

1.1
The Disperse Particles

The primary pigment particles (normally in the submicron range) are responsible for the opacity, color and anti-corrosive properties. The principal pigment in use is titanium dioxide due to its high refractive index, and is the one that is used to produce white paint. To produce maximum scattering, the particle size distribution of titanium dioxide has to be controlled within a narrow limit. Rutile, with a refractive index of 2.76, is preferred over anatase, which has a lower refractive index of 2.55. Thus, rutile gives the possibility of higher opacity than anatase and it is more resistant to chalking on exterior exposure. To obtain maximum opacity, the particle size of rutile should be within the range 220–140 nm. The surface of rutile is photoactive and it is surface coated with silica and alumina in various proportions to reduce its photoactivity.

Colored pigments may consist of inorganic or organic particles. For a black pigment, one can use carbon black, copper carbonate, manganese dioxide (inorganic) and aniline black (organic). For yellow, can use lead, zinc, chromates, cadmium sulfide, iron oxides (inorganic) and nickel azo yellow (organic). For blue–violet, one can use ultramarine, Prussian blue, cobalt blue (inorganic), phthalocyanine, indanthrone blue and carbazole violet (organic). For red, one can use red iron oxide, cadmium selenide, red lead, chrome red (inorganic), toluidine red and quinacridones (organic).

The color of a pigment is determined by the selective absorption and reflection of the various wavelengths of visible light (400–700 nm) which impinges on it. For example, a blue pigment appears so because it reflects the blue wavelengths in the incident white light and absorbs the other wavelengths. Black pigments absorb all the wavelengths of incident light almost totally, whereas a white pigment reflects all the visible wavelengths.

The primary shape of a pigmented particle is determined by its chemical nature, its crystalline structure (or lack of it) and the way in which the pigment is created in Nature or made synthetically. Pigments as primary particles may be spherical, nodular, needle- or rod-like or plate like (lamellar) as illustrated in Figure 1.1.

The pigments are usually supplied in the form of aggregates (whereby the particles are attached at their faces) or agglomerates (where the particles are attached at their corners). When dispersed in the continuous phase, these aggregates and agglomerates must be dispersed into single units. This requires the use of an effective wetter/dispersant and application of mechanical energy. This process of dispersion is discussed in detail in Chapter 3.

In paint formulations, secondary pigments are also used. These are referred to as extenders, fillers and supplementary pigments. They are relatively cheaper than the primary pigments and they are incorporated in conjunction with the primary pigments for a variety of reasons, such as cost effectiveness, enhancement of adhesion, reduction of water permeability and enhancement of corrosion resistance. For example, in a primer or undercoat (matt latex paint), coarse particle extenders such as calcium carbonate are added in conjunction with TiO_2 to achieve whiteness and opacity in a matt or semi-matt product. The particle size of extenders range from submicron to a few tens of microns.

Their refractive index is very close to that of the binder and hence they do not contribute to the opacity from light scattering. Most extenders used in the paint industry are naturally occurring materials such as barytes (barium sulfate), chalk (calcium carbonate), gypsum (calcium sulfate) and silicates (silica, clay, talc or mica). However, more recently synthetic polymeric extenders have been designed to replace some of the TiO_2. A good example is spindrift, which is polymer beads that consist of spherical particles (up to 30 μm in diameter) which contain submicron air bubbles and a small proportion of TiO_2. The small air bubbles (<0.1 μm) reduce the effective refractive index of the polymer matrix, thus enhancing the light scattering of TiO_2.

The refractive index of any material (primary or secondary pigment) is a key to its performance. As is well known, the larger the difference in refractive index between the pigment and the medium in which it is dispersed, the greater is the opacity effect. A summary of the refractive indices of various extender and opacifying pigments is given in Table 1.1.

The refractive index of the medium in which the pigment is dispersed ranges from 1.33 (for water) to 1.4–1.6 (for most film formers). Thus rutile will give the highest opacity, whereas talc and calcium carbonate will be transparent in fully bound surface coatings. Another important fact that affects light scattering is the particle size, and to obtain the maximum opacity from rutile an optimum particle

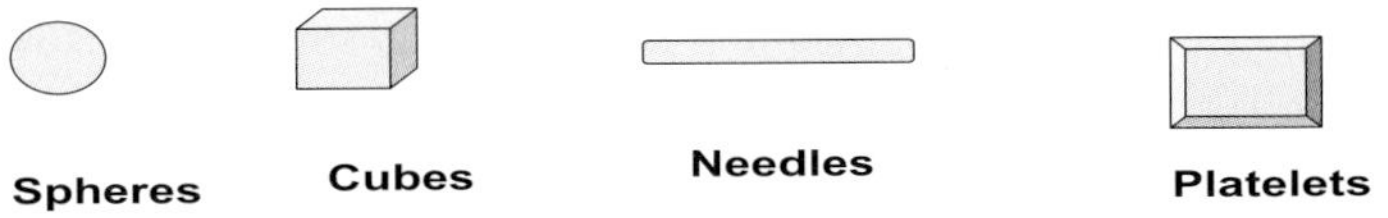

Figure 1.1 Schematic representation of particle shape.

Table 1.1 Refractive indices (RI) of extenders and opacifying pigments.

Extender pigment	RI	Opacifying white pigment	RI
Calcium carbonate	1.58	Zinc sulfide/Barium sulfate	1.84
China clay	1.56	Zinc oxide	2.01
Talc	1.55	Zinc sulfide	2.37
Barytes	1.64	TiO_2 anatase	2.55
		TiO_2 rutile	2.76

size of 250 nm is required. This explains the importance of good dispersion the powder in the liquid, which can be achieved by a good wetting/dispersing agent and also application of sufficient milling efficiency.

For colored pigments, the refractive index of a pigment in the non-absorbing or highly reflecting part of the spectrum affects its performance as an opacifying material. For example, Pigment Yellow 1 and Arylamide Yellow G give lower opacity than Pigment Yellow 34 Lead Chromate. Most suppliers of colored pigments attempt to increase the opacifying effect by controlling the particle size.

The nature of the pigment surface plays a very important role in its dispersion in the medium and its affinity to the binder. For example, the polarity of the pigment determines its affinity for alkyds, polyesters, acrylic polymers and latexes that are commonly used as film formers (see below). In addition, the nature of the pigment surface determines its wetting characteristics in the medium in which it is dispersed (which can be aqueous or non-aqueous) and also the dispersion of the aggregates and agglomerates into single particles. It also affects the overall stability of the liquid paint. Most pigments are surface treated by the manufacturer to achieve the optimum performance. As mentioned above, the surface of rutile particles is treated with silica and alumina in various proportions to reduce its photoactivity. If the pigment has to be used in a non-aqueous paint, its surface is also treated with fatty acids and amines to make it hydrophobic for incorporation in an organic medium. This surface treatment enhances the dispersibility of the paint, its opacity and tinting strength and its durability (glass retention, resistance to chalking and color retention). It can also protect the binder in the paint formulation.

The dispersion of the pigment powder in the continuous medium requires several processes, namely wetting of the external and internal surface of the aggregates and agglomerates, separation of the particles from these aggregates and agglomerates by application of mechanical energy, displacement of occluded air and coating of the particles with the dispersion resin. It is also necessary to stabilize the particles against flocculation by either electrostatic double-layer repulsion and/or steric repulsion. The process of wetting and dispersion of pigments is described in detail in Chapter 3 and colloid stability (lack of aggregation) is discussed in Chapter 4.

1.2 The Dispersion Medium and Film Formers

The dispersion medium can be aqueous or non-aqueous, depending on the application. It consists of a dispersion of the binder in the liquid (which is sometimes referred to as the diluent). The term solvent is frequently used to include liquids that do not dissolve the polymeric binder. Solvents are used in paints to enable the paint to be made and they allow application of the paint to the surface. In most cases, the solvent is removed after application by simple evaporation, and if the solvent is completely removed from the paint film it should not affect the paint film performance. However, in the early life of the film solvent retention, it can affect hardness, flexibility and other film properties. In water-based paints, the water may act as a true solvent for some of the components, but it should be a non-solvent for the film former. This is particularly the case with emulsion paints.

With the exception of water, all solvents, diluents and thinners used in surface coatings are organic liquids with low molecular weight. Two types can be distinguished, hydrocarbons (both aliphatic and aromatic) and oxygenated compounds such as ethers, ketones, esters and ether alcohols. Solvents, thinners and diluents control the flow of the wet paint on the substrate to achieve a satisfactory smooth, even, thin film, which dries in a predetermined time. In most cases, mixtures of solvents are used to obtain the optimum condition for paint application. The main factors that must be considered when choosing solvent mixtures are their solvency, viscosity, boiling point, evaporation rate, flash point, chemical nature, odor and toxicity.

The solvent power or solvency of a given liquid or mixture of liquids determines the miscibility of the polymer binder or resin. It has also a large effect on the attraction between particles in a paint formulation, as discussed in detail in Chapter 4. A very useful parameter that describes solvency is the Hildebrand solubility parameter, δ [2, 3], which is related to the energy of association of molecules in the liquid phase in terms of 'cohesive energy density'. The latter is simply the ratio of the energy required to vaporize 1 cm^3 of liquid, ΔE_v, to its molar volume, V_m. The solubility parameter δ is simply the square root of that ratio,

$$\delta = \frac{\Delta E_v}{V_m} \quad (1)$$

Liquids having similar values of δ are miscible, whereas those with a significant difference are immiscible. The solubility parameters of liquids can be determined experimentally by measuring the energy of vaporization. For polymers, one can determine the solubility parameter using an empirical approach by contacting the polymer with liquids with various δ values and observing whether or not dissolution occurs. The solubility parameter of the polymer is taken the average of two δ values for two solvents that appear to dissolve the polymer. A better method is to calculate the solubility parameter from the 'molar attraction constant', G, of the constituent parts of the molecule [4]:

$$\delta = \frac{\rho \Sigma G}{M} \tag{2}$$

where ρ is the density of the polymer and M is its molecular weight.

Hansen [5] extended Hildebrand's concept by considering three components for the solubility parameter, a dispersion component δ_{d}, a polar component δ_{p} and a hydrogen bonding component δ_{h}:

$$\delta^2 = {\delta_{\mathrm{d}}}^2 + {\delta_{\mathrm{p}}}^2 + {\delta_{\mathrm{h}}}^2 \tag{3}$$

Values of δ and its components are tabulated in the book by Barton [4].

As mentioned above, the dispersion medium consist of a solvent or diluent and the film former. The latter is also sometimes referred to as a 'binder', since it functions by binding the particulate components together, and this provides the continuous film-forming portion of the coating. The film former can be a low molecular weight polymer (oleoresinous binder, alkyd, polyurethane, amino resin, epoxide resin, unsaturated polyester), a high molecular weight polymer (nitrocellulose, solution vinyls, solution acrylics), an aqueous latex dispersion [poly(vinyl acetate), acrylic, styrene–butadiene] or a non-aqueous polymer dispersion (NAD). This introductory chapter only briefly describes film formers based on polymer solutions. The subject of polymer latexes and non-aqueous dispersions is dealt with in Chapter 2. The polymer solution may exist in the form of a fine particle dispersion in a non-solvent. In some cases, the system may be a mixed solution–dispersion, implying that the solution contains both single polymer chains and aggregates of these chains (sometimes referred to as micelles). A striking difference between a polymer that is completely soluble in the medium and one that contains aggregates of that polymer is the viscosity reached in the two cases. A polymer that is completely soluble in the medium will show a higher viscosity at a given concentration compared with another polymer (at the same concentration) that produces aggregates. Another important difference is the rapid increase in the solution viscosity with increase in molecular weight for a completely soluble polymer. If the polymer makes aggregates in solution, an increase in the molecular weight of the polymer does not give a dramatic increase in viscosity.

The earliest film-forming polymers used in paints were based on natural oils, gums and resins. Modified natural products are based on cellulose derivatives such as nitrocellulose, which is obtained by nitration of cellulose under carefully specified conditions. Organic esters of cellulose such as acetate and butyrate can also be produced. Another class of naturally occurring film formers are those based on vegetable oils and their derived fatty acids (renewable resource materials). Oils used in coatings include linseed oil, soya bean oil, coconut oil and tall oil. When chemically combined into resins, the oil contributes flexibility and, with many oils, oxidative cross-linking potential. The oil can also be chemically modified, as for example by the hydrogenation of castor oil, which can be combined with alkyd resins to produce some specific properties of the coating.

Another early binder used in paints is the oleoresinous vehicles produced by heating together oils and either natural or certain preformed resins, so that the resin dissolves or disperses in the oil portion of the vehicle. However, these oleoresinous vehicles were later replaced by alkyd resins, which probably constitute of the first applications of synthetic polymers in the coating industry. These alkyd resins are polyesters obtained by reaction of vegetable oil triglycerides, polyols (e.g. glycerol) and dibasic acids or their anhydrides. These alkyd resins enhance the mechanical strength, drying speed and durability over and above those obtained using the oleoresinous vehicles. The alkyds were also modified by replacing part of the dibasic acid with a diisocyanate (such as toluene diisocyanate) to produce greater toughness and quicker drying characteristics.

Another type of binder is based on polyester resins (both saturated and unsaturated). These are typically composed mainly of co-reacted di- or polyhydric alcohols and di- or tribasic acids or acid anhydrides. They have also been modified using silicone to enhance their durability.

More recently, acrylic polymers have been used in paints due to their excellent properties of clarity, strength and chemical and weather resistance. Acrylic polymers refer to systems containing acrylate and methylacrylate esters in their structure along with other vinyl unsaturated compounds. Both thermoplastic and thermosetting systems can be made; the latter are formulated to include monomers possessing additional functional groups that can further react to give cross-links following the formation of the initial polymer structure. These acrylic polymers are synthesized by radical polymerization. The main polymer-forming reaction is a chain propagation step which follows an initial initiation process. A variety of chain transfer reactions are possible before chain growth ceases by a termination process.

Radicals produced by transfer, if sufficiently active, can initiate new polymer chains where a monomer is present that is readily polymerized. Radicals produced by chain transfer agents (low molecular weight thiols, e.g. primary octanethiol) are designed to initiate new polymer chains. These agents are introduced to control the molecular weight of the polymer.

The monomers used for preparation of acrylic polymers vary in nature and can generally be classified as 'hard' (such as methyl methacrylate, styrene and vinyl acetate) or 'soft' (such as ethyl acrylate, butyl acrylate and 2-ethylhexyl acrylate). Reactive monomers may also have hydroxyl groups (such as hydroxyethyl acrylate). Acidic monomers such as methacrylic acid are also reactive and may be included in small amounts in order that the acid groups may enhance pigment dispersion. The practical coating systems are usually copolymers of 'hard' and 'soft'. The polymer hardness is characterized by its glass transition temperature, T_g. The T_g(K) of the copolymer can be estimated from the T_g values of the individual components of the homopolymers with weight fractions W_1 and W_2:

$$\frac{1}{T_g} = \frac{W_1}{T_{g1}} + \frac{W_2}{T_{g2}} \tag{4}$$

Most acrylic polymers consist of random copolymers. By controlling the proportion of 'hard' and 'soft' monomers and the molecular weight of the final copolymer, one arrives at the right property that is required for a given coating. As mentioned above, two types of acrylic resins can be produced, namely thermoplastic and thermosetting. The former find application in automotive topcoats, although they suffer from some disadvantages such as cracking in cold conditions, which may require a process of plasticization. These problems are overcome by using thermosetting acrylics, which improve the chemical and alkali resistance. Also, it allows one to use higher solid contents in cheaper solvents. Thermosetting resins can be self-cross-linking or may require a co-reacting polymer or hardener.

1.3
Deposition of Particles and Their Adhesion to the Substrate

In a paint film, the pigment particles need to undergo a process of deposition to the surfaces (which is governed by long-range forces such as van der Waals attraction and electrical double-layer repulsion or attraction). This process of deposition is also affected by polymers (nonionic, anionic or cationic) which can enhance or prevent adhesion. Once the particles reach the surface, they have to adhere strongly to the substrate. This process of adhesion is governed by short-range forces (chemical or non-chemical). The same applies to latex particles, which also undergo a process of deposition, adhesion and coalescence. The subject of particle deposition and adhesion is discussed in detail in Chapter 5.

1.4
Flow Characteristics (Rheology) of Paints

Control of the flow characteristics of paints is essential for their successful application. All paints are complex systems consisting of various components such as pigments, film formers, latexes and rheology modifiers. These components interact with each other and the final formulation becomes non-Newtonian, showing complex rheological behavior. The paint is usually applied in three stages, namely transfer of the paint from the bulk container, transfer of the paint from the applicator (brush or roller) to the surface to form a thin, even film and flow-out of film surface, coalescence of polymer particles (latexes) and loss of the medium by evaporation. During each of these processes, the flow characteristics of the paint and its time relaxation produce interesting rheological responses. To understand the rheological behavior of a paint system, one must start with a basic knowledge of rheology. This is described in Chapter 6, which gives an account of the basic principles of rheology. Chapter 7 describes the rheology of suspensions and emulsions, which are main components of the paint formulation. Chapter 8 is devoted to the subject of rheology modifiers, which are essential components of a paint formulation. These modifiers are introduced for various reasons, such as maintenance of the long-term stability of the paint, ease of its application and the final characteristics of the film produced. The final chapter gives a brief account of the main rheological characteristics of a paint formulation.

References

1 R. Lambourne (ed.), *Paint and Surface Coatings: Theory and Practice*, Ellis Horwood, Chichester (1987).
2 J.H. Hildebrand and R. Scott, *The Solubility of Non-electrolytes*, 3rd edn, Reinhold, New York (1950).
3 J.H. Hildebrand and R. Scott, *Regular Solutions*, Prentice Hall, Englewood Cliffs, NJ (1962).
4 A.F.M. Barton, *Handbook of Solubility Parameters*, CRC Press, New York (1983).
5 C.M. Hansen, *J. Paint Technol.*, *39* (505), 104 (1967); *39* (511), 505 (1967).

2
Emulsion, Dispersion and Suspension Polymerization: Preparation of Polymer Colloids and Their Stabilization

Emulsion polymers (latexes) are the most commonly used film formers in the coating industry. This is particularly the case with aqueous emulsion paints that are used for home decoration. These aqueous emulsion paints are applied at room temperature and the latexes coalesce on the substrate, forming a thermoplastic film. Sometimes functional polymers are used for cross-linking in the coating system. The polymer particles are typically submicron (0.1–0.5 μm).

Generally, there are three methods for the preparation of polymer dispersions, namely emulsion, dispersion and suspension polymerization [1]. In emulsion polymerization, a monomer is emulsified in a non-solvent, commonly water, usually in the presence of a surfactant. A water-soluble initiator is added and particles of polymer form and grow in the aqueous medium as the reservoir of the monomer in the emulsified droplets is gradually used up. In dispersion polymerization [which is usually applied for the preparation of non-aqueous polymer dispersions, commonly referred to as non-aqueous dispersion (NAD) polymerization], monomer, initiator, stabilizer (referred to as a protective agent) and solvent initially form a homogeneous solution [2]. The polymer particles precipitate when the solubility limit of the polymer is exceeded. The particles continue to grow until the monomer is consumed. In suspension polymerization, the monomer is emulsified in the continuous phase using a surfactant or polymeric suspending agent. The initiator (which is oil soluble) is dissolved in the monomer droplets and the droplets are converted into insoluble particles, but no new particles are formed.

Below, a description of both emulsion and dispersion polymerization is given, with particular reference to the control of their particle size and colloid stability, which is greatly influenced by the emulsifier or dispersant used. Particular emphasis is given to the effect of polymeric surfactants that has been recently used for the preparation of emulsion polymers.

2.1
Emulsion Polymerization

As mentioned above, in emulsion polymerization, the monomer, e.g. styrene or methyl methacrylate, that is insoluble in the continuous phase, is emulsified

Colloids and Interface Science Series, Vol. 6
Colloids in Paints. Tharwat F. Tadros

ISBN: 978-3-527-31466-9

using a surfactant that adsorbs at the monomer/water interface [1]. The surfactant micelles in bulk solution solubilizes some of the monomer. A water-soluble initiator such as potassium persulfate ($K_2S_2O_8$) is added and this decomposes in the aqueous phase, forming free radicals that interact with the monomers forming oligomeric chains. It has long been assumed that nucleation occurs in the 'monomer-swollen micelles' [3]. The reasoning behind this mechanism was the sharp increase in the rate of reaction above the critical micelle concentration and that the number of particles formed and their size depend to a large extent on the nature of the surfactant and its concentration (which determines the number of micelles formed). However, later this mechanism was disputed and it was suggested that the presence of micelles means that excess surfactant is available and molecules will readily diffuse to any interface.

The most accepted theory of emulsion polymerization is referred to as the coagulative nucleation theory [4, 5]. A two-step coagulative nucleation model has been proposed by Napper and co-workers [4, 5]. In this process, the oligomers grow by propagation and this is followed by a termination process in the continuous phase. A random coil is produced which is insoluble in the medium, and this produces a precursor oligomer at the θ-point. The precursor particles subsequently grow primarily by coagulation to form true latex particles. Some growth may also occur by further polymerization. The colloidal instability of the precursor particles may arise from their small size and the slow rate of polymerization can be due to reduced swelling of the particles by the hydrophilic monomer [4, 5]. The role of surfactants in these processes is crucial, since they determine the stabilizing efficiency and the effectiveness of the surface-active agent ultimately determines the number of particles formed. This was confirmed by using surface-active agents of different nature. The effectiveness of any surface-active agent in stabilizing the particles was the dominant factor and the number of micelles formed was relatively unimportant.

A typical emulsion polymerization formulation contains water, 50% monomer blended for the required T_g (see Chapter 1), surfactant (and often colloid), initiator, pH buffer and fungicide. Hard monomers with a high T_g used in emulsion polymerization include vinyl acetate, methyl methacrylate and styrene. Soft monomers with a low T_g include butyl acrylate, 2-ethylhexyl acrylate, vinyl versatate and maleate esters. The most suitable monomers are those with low, but not too low, water solubility. Other monomers, such as acrylic acid, methacrylic acid and adhesion promoting monomers, may be included in the formulation. It is important that the latex particles coalesce as the diluent evaporates. The minimum film-forming temperature (MFFT) of the paint is a characteristic of the paint system. It is closely related to the T_g of the polymer, but the latter can be affected by materials present such as surfactant and the inhomogeneity of the polymer composition at the surface. High-T_g polymers will not coalesce at room temperature and in this case a plasticizer ('coalescing agent') such as benzyl alcohol is incorporated in the formulation to reduce the T_g of the polymer, thus reducing the MFFT of the paint. Clearly, for any paint system one must determine the MFFT since, as mentioned above, the T_g of the polymer is greatly affected by the ingredients in the paint formulation.

Several types of surfactants can be used in emulsion polymerization, and a summary of the various classes is given in Table 2.1.

The role of surfactants is twofold, first to provide a locus for the monomer to polymerize and second to stabilize the polymer particles as they form. In addition, surfactants aggregate to form micelles (above the critical micelle concentration), and these can solubilize the monomers. In most cases, a mixture of anionic and nonionic surfactant is used for optimum preparation of polymer latexes. Cationic surfactants are seldom used, except for some specific applications where a positive charge is required on the surface of the polymer particles.

In addition to surfactants, most latex preparations require the addition of a polymer (sometimes referred to as a 'protective colloid'), such as partially hydrolyzed poly(vinyl acetate) [commercially referred to as poly(vinyl alcohol) (PVA)], hydroxyethylcellulose or a block copolymer of poly(ethylene oxide) (PEO) and poly(propylene oxide) (PPO). These polymers can be supplied with various molecular weights or proportions of PEO and PPO. When used in emulsion polymerization, they can be grafted by the growing chain of the polymer being formed. They assist in controlling the particle size of the latex, enhancing the stability of the polymer dispersion and controlling the rheology of the final paint.

Table 2.1 Surfactants used in emulsion polymerization.

Type	Surfactants
Anionic	Carboxylates: C_nH_{2n+1} COO^-X Sulfates: C_nH_{2n+1} OSO_3^-X Sulfonates: C_nH_{2n+1} SO_3^-X Phosphates: C_nH_{2n+1} $OPO(OH)O^-X$ with n being the range 8–16 atoms and the counterion X is usually Na^+ Several other anionic surfactants are commercially available, such as sulfosuccinates, isothionates and taurates, and these are sometimes used for special applications.
Cationic	Alkyltrimethylammonium chloride, $R(CH_3)_3NCl$, where R contains 8–18 C atoms, e.g. dodecyltrimethylammonium chloride, $C_{12}H_{25}(CH_3)_3NCl$
Zwitterions	*N*-Alkylbetaines, which are derivatives of trimethylglycine $(CH_3)_3NCH_2COOH$ (betaine). An example of betaine surfactants is laurylamidopropyldimethylbetaine, $C_{12}H_{25}CON(CH_3)_2CH_2COOH$. These alkylbetaines are sometimes described as alkyl dimethylglycinates
Nonionic	Alcohol ethoxylates, alkylphenol ethoxylates, fatty acid ethoxylates, monoalkanolamide ethoxylates, sorbitan ester ethoxylates, fatty amine ethoxylates and ethylene oxide–propylene oxide copolymers (sometimes referred to as polymeric surfactants). Multihydroxy products such as glycol esters, glycerol (and polyglycerol) esters, glucosides (and polyglucosides), sucrose esters. Amine oxides and sulfinyl surfactants (nonionics with a small head group).

A typical emulsion polymerization process involves two stages, known as the seed stage and the feed stage. In the feed stage, an aqueous charge of water, surfactant and colloid is raised to the reaction temperature (85–90 °C) and 5–10% of the monomer mixture is added along with a proportion of the initiator (a water-soluble persulfate). In this seed stage, the formulation contains monomer droplets stabilized by surfactant, a small amount of monomer in solution and surfactant monomers and micelles. Radicals are formed in solution from the breakdown of the initiator and these radicals polymerize the small amount of monomer in solution. These oligomeric chains will grow to some critical size, the length of which depends on the solubility of the monomer in water. The oligomers build up to a limiting concentration and this is followed by a precipitous formation of aggregates (seeds), a process similar to micelle formation, except in this case the aggregation process is irreversible (unlike surfactant micelles, which are in dynamic equilibrium with monomers).

In the feed stage, the remaining monomer and initiator are fed together and the monomer droplets become emulsified by the surfactant remaining in solution (or by extra addition of surfactant). Polymerization proceeds as the monomer diffuses from the droplets, through the water phase, into the already forming growing particles. At the same time, radicals enter the monomer-swollen particles, causing both termination and re-initiation of polymerization. As the particles grow, the remaining surfactant from the water phase is adsorbed on the surface of particles to stabilize the polymer particles. The stabilization mechanism involves both electrostatic and steric repulsion, as discussed in Chapter 4. The final stage of polymerization may include a further shot of initiator to complete the conversion.

2.1.1
Mechanism of Emulsion Polymerization

According to the theory of Smith and Ewart [3] of the kinetics of emulsion polymerization, the rate of propagation R_p is related to the number of particles N formed in a reaction by the equation

$$-\frac{d[M]}{dt} = R_p k_p N n_{av}[M] \tag{1}$$

where [M] is the monomer concentration in the particles, k_p is the propagation rate constant and n_{av} is the average number of radicals per particle.

According to equation 1, the rate of polymerization and the number of particles are directly related to each other, i.e. an increase in the number of particles will increase the rate. This has been found for many polymerizations, although here are some exceptions. The number of particles is related to the surfactant concentration [S] by the equation

$$N \approx [S]^{\frac{3}{5}} \tag{2}$$

Using the coagulative nucleation model, Napper and co-workers [4, 5] found that the final particle number increases with increase in surfactant concentration with a monotonically diminishing exponent. The slope of d(log N_c)/d(log t) varies from 0.4 to 1.2. At high surfactant concentration, the nucleation time will be long since the new precursor particles will be readily stabilized. As a result, more latex particles are formed and eventually will outnumber the very small precursor particles at long times. The precursor–particle collisions will become more frequent and fewer latex particles are produced. dN_c/dt will approach zero and at long times the number of latex particles will remain constant. This shows the inadequacy of the Smith–Ewart theory, which predicts a constant exponent (3/5) at all surfactant concentrations. For this reason, the coagulative nucleation mechanism has now been accepted as the most probable theory for emulsion polymerization. In all cases, the nature and concentration of surfactant used are very crucial and this is very important in the industrial preparation of latex systems.

Most reports on emulsion polymerization have been limited to commercially available surfactants, which in many cases are relatively simple molecules such as sodium dodecyl sulfate, and simple nonionic surfactants. However, studies on the effect of surfactant structure on latex formation have revealed the importance of the structure of the molecule. Block and graft copolymers (polymeric surfactants) are expected to be better stabilizers than simple surfactants. The use of these polymeric surfactants in emulsion polymerization and the stabilization of the resulting polymer particles are discussed below.

2.1.2 Block Copolymers as Stabilizers in Emulsion Polymerization

Most aqueous emulsion and dispersion polymerizations reported in the literature are based on a few commercial products with a broad molecular weight distribution and varying block composition. The results obtained from these studies could not establish what effect the structural features of the block copolymer have on their stabilizing ability and effectiveness in polymerization. Fortunately, model block copolymers with well-defined structures could be synthesized and their role in emulsion polymerization has been studied using model polymers and model latexes.

A series of well-defined A–B block copolymers of polystyrene-block-poly(ethylene oxide) (PS–PEO) were synthesized [6] and used for emulsion polymerization of styrene. These molecules are 'ideal' since the polystyrene block is compatible with the polystyrene formed and thus it forms the best anchor chain. The PEO chain (the stabilizing chain) is strongly hydrated with water molecules and it extends into the aqueous phase, forming the steric layer necessary for stabilization (see Chapter 4). However, the PEO chain can become dehydrated at high temperature (due to the cleavage of hydrogen bonds), thus reducing the effective steric stabilization. Hence the emulsion polymerization should be carried out at temperatures well below the theta (θ) temperature of PEO (see Chapter 4).

Five block copolymers were synthesized [6] with various molecular weights of the PS and PEO blocks. The molecular weight of the polystyrene block and the resulting PS–PEO polymer was determined using gel permeation chromatography. The molar percentage of ethylene oxide and the percentage of PEO in the block were determined using ^{1}H NMR spectroscopy. The molecular weight of the blocks varied from $M_n = 1000$ to 7000 for PS and from $M_w = 3000$ to 9000 for PEO. These five block copolymers were used for emulsion polymerization of styrene at 50 °C (well below the θ-temperature of PEO). The results indicated that for efficient anchoring, the PS block need not be more than 10 monomer units. The PEO block should have an $M_w \geq 3000$. However, the ratio of the two blocks is very important; for example, if the concentration of PEO is ≤ 3000 wt%, the molecule becomes insoluble in water (not sufficiently hydrophilic) and no polymerization could occur when using this block copolymer. In addition, the 50% PEO block could produce a latex but the latter was unstable and coagulated at 35% conversion. It became clear from these studies that the percentage PEO in the block copolymer plays an important role and it should exceed 75%. However, the overall molecular weight of the block copolymer is also very important. For example if one uses a PS block with $M_n = 7000$, the PEO molecular weight has to be 21 000, which is too high and may result in bridging flocculation, unless one prepares a very dilute latex.

Another systematic study of the effect of block copolymer on emulsion polymerization was carried out using block of poly(methyl methacrylate)-block-poly(ethylene oxide) (PMMA–PEO) for the preparation of PMMA latexes [6]. The ratio and molecular weight of PMMA to PEO in the block copolymer were varied. Ten different PMMA–PEO blocks were synthesized [6] with M_n for PMMA varying between 400 and 2500. The M_w of PEO was varied between 750 and 5000. The recipe for MMA polymerization consisted of 100 g of monomer, 800 g of water, 20 g of PMMA–PEO block copolymer and 0.5 g of potassium persulfate. The polymerization was carried out at 45 °C, which is well below the θ-temperature of PEO. The rate of polymerization, R_p, was calculating by using latex samples drawn from the reaction mixture at various intervals of time (the amount of latex was determined gravimetrically). The particle size of each latex was determined by dynamic light scattering [photon correlation spectroscopy (PCS)]. The number of particles N in each case was calculated from the weight of the latex and the z-average diameter. The results obtained were used to study the effect of the anchoring group PMMA molecular weight, the effect of PEO molecular weight and the effect of the total molecular weight of the block copolymer. The results are summarized in Tables 2.2 and 2.3.

These results of the systematic study of varying the PMMA and PEO block molecular weights, the percentage of PEO in the chain and the overall molecular weight clearly show the effect of these factors on the resulting latex. For example, when using a block copolymer with a molecular weight of PMMA of 400 and a molecular weight of PEO of 750 (i.e. containing 65 wt% PEO), the resulting latex had fewer particles compared with the other surfactants. The most dramatic effect was obtained when the PMMA molecular weight was increased to 900 while

Table 2.2 Effect of PMMA and PEO molecular weight in the diblock.

M_n PMMA	M_w PEO	wt% PEO	$R_p \times 10^4$/mol l^{-1} s^{-1}	D/nm	$N \times 10^{-13}$/cm^{-3}
400	750	65	1.3	213	1.7
400	2000	83	1.5	103	14.7
400	5000	93	2.4	116	10.3
900	750	46	Unstable latex	–	–
800	2000	71	3.4	92	20.6
800	5000	86	3.2	106	13.5
1300	2000	61	2.4	116	10.3
1200	5000	81	4.6	99	16.6
1900	5000	72	3.4	110	11.4
2500	5000	67	2.2	322	0.4

Table 2.3 Effect of total molecular weight of the PMMA–PEO diblock.

M_w	wt% PEO	$R_p \times 10^4$/mol l^{-1} s^{-1}	D/nm	$N \times 10^{-13}$/cm^{-3}
1150	65	1.3	213	1.7
2400	83	1.5	103	14.7
2800	71	3.4	92	20.6
3300	61	2.4	99	16.6
6200	81	4.6	99	16.6
6900	72	3.4	110	11.4
7500	67	2.2	322	0.4

keeping the PEO molecular weight the same (750). This block copolymer contained only 46 wt% PEO and it became insoluble in water due to the lack of hydrophilicity. The latex produced was unstable and it collapsed at an early stage of polymerization. The PEO molecular weight of 750 is insufficient to provide effective steric stabilization (see Chapter 4). By increasing the molecular weight of PEO to 2000 or 5000 while keeping the PMMA molecular weight at 400 or 800, a stable latex was produced with a small particle diameter and large number of particles. The best results were obtained by keeping the molecular weight of PMMA at 800 and that of PEO at 2000. This block copolymer gave the highest conversion rate, the smallest particle diameter and the largest number of particles (see Table 2.3). It is interesting that on increasing the PEO molecular weight to 5000 while keeping the PMMA molecular weight at 800, the rate of conversion decreased, the average diameter increased and the number of particles decreased compared with the results obtained using a molecular weight of 2000 for PEO. It seems that when the PEO molecular weight was increased, the hydrophilicity of the molecule increased (86 wt% PEO), and this reduced the efficiency of the copolymer. It seems that by increasing the hydrophilicity of the block copolymer and its overall molecular weight, the rate of adsorption of the polymer to the latex particles and its overall adsorption strength may have decreased. The effect of the overall molecular weight of the block copolymer and its overall hydrophilicity have a large effect on the latex production (see Table 2.3). Increasing the overall molecular weight of the block copolymer to >6200 resulted in a reduction in the rate of conversion, an increase in the particle diameter and a reduction in the number of latex particles. The worst results were obtained with an overall molecular weight of 7500 while reducing the percentage of PEO, in which case particles with 322 nm diameter were obtained and the number of latex particles was significantly reduced.

The importance of the affinity of the anchor chain (PMMA) to the latex particles was investigated by using different monomers [6]. For example, when using styrene as the monomer, the resulting latex was unstable and it showed the presence of coagulum. This can be attributed to the lack of chemical compatibility of the anchor chain (PMMA) and the polymer to be stabilized, namely polystyrene. This clearly indicates that PMMA–PEO block copolymers are not suitable for emulsion polymerization of styrene. However, when using vinyl acetate monomer, whereby the resulting poly(vinyl acetate) latex should have a strong affinity to the PMMA anchor, no latex was produced when the reaction was carried out at 45 °C. It was speculated that the water solubility of the vinyl acetate monomer resulted in the formation of oligomeric chain radicals which could exist in solution without nucleation. Polymerization at 60 °C, which did nucleate particles, was found to be controlled by chain transfer of the vinyl acetate radical with the surfactant, resulting in broad molecular weight distributions.

Emulsion polymerization of MMA using triblock copolymers of PMMA-block-PEO–PMMA were carried out using blocks with the same PMMA molecular weight (800 or 900) while varying the PEO molecular weight from 3400 to 14 000 in order to vary the loop size. Although the rate of polymerization was not affected by the loop size, the particles with the smallest diameter were obtained with the 10 000 molecular weight PEO. Comparison of the results obtained using the triblock copolymer with those obtained using the diblock copolymer (while keeping the PMMA block molecular weight the same) showed the same rate of polymerization. However, the average particle diameter was smaller and the total number of particles was larger when using the diblock copolymer. This clearly shows the higher efficacy of the diblock copolymer than the triblock copolymer.

2.1.3
Graft Copolymers as Stabilizers in Emulsion Polymerization

The first systematic study of the effect of graft copolymers were carried out by Piirma and Lenzotti [7], who synthesized well-characterized graft copolymers with different backbone and side-chain lengths. Several grafts of poly(*p*-

methylstyrene)-graft-poly(ethylene oxide), PMSt(PEO)$_n$, were synthesized and used in styrene emulsion polymerization. Three different PMSt chain lengths (with molecular weights of 750, 2000 and 5000) and three different PEO chain lengths were prepared. In this way, the structure of the amphipathic graft copolymer could be changed in three different ways: (i) three different PEO graft chain lengths; (ii) three different backbone chain lengths with the same weight-percent of PEO; and (iv) four different weight-percent PEO grafts.

Piirma and Lenzotti [7] first investigated the graft copolymer concentration required to produce the highest conversion rate, the smallest particle size and the largest number of latex particles. The monomer-to-water ratio was kept at 0.15 to avoid overcrowding of the resulting particles. They found that a concentration of 18 g per 100 g of monomer (2.7% aqueous phase) was necessary to obtain the above results, after which further increases in graft copolymer concentration did not result in any significant increase in the rate of polymerization or increase in the number of particles used. Using the graft copolymer concentration of 2.7% aqueous phase, the results showed an increase in the number of particles with increase in conversion, reaching a steady value at about 35% conversion. Obviously, before that conversion, new particles are still being stabilized from the oligomeric precursor particles, after which all precursor particles are assimilated by the existing particles. The small size of the latex produced, namely 30–40 nm, clearly indicates the efficiency with which this graft copolymer stabilizes the dispersion.

Three different backbone chain lengths of M_n 1140, 4270 and 24 000 were studied while keeping the same weight-percentage of PEO (82%), which are equivalent to 3, 10 and 55 PEO chains per backbone, respectively. The results showed that the rate of polymerization, particle diameter and number of particles were similar for the three cases. Since the graft copolymer concentration was the same in each case, it can be concluded that one molecule of the highest molecular graft is just as effective as 18 molecules of the lowest molecular weight graft in stabilizing the particles.

Four graft copolymers were synthesized with a PMSt backbone with $M_w = 4540$ while increasing the PEO content: 68, 73, 82 and 92 wt% (corresponding to 4.8, 6, 10 and 36 grafts per chain, respectively). The results showed a sharp decrease (by more than one order of magnitude) in the number of particles as the weight-percentage of PEO is increased from 82 to 94%. The reason for this reduction in the number of particles is the increased hydrophilicity of the graft copolymer, which could result in desorption of the molecule from the surface of the particle. In addition, a graft with 36 side chains doest not leave enough space for anchoring by the backbone.

The effect of PEO side-chain length on emulsion polymerization using graft copolymers was systematically studied by keeping the backbone molecular weight the same (1380) while gradually increasing the PEO molecular weight of the side chains from 750 to 5000. For example, by increasing M_w of PEO from 750 to 2000 while keeping the PEO content roughly the same (84 and 82 wt%, respectively), the number of side chains in the graft decreased from 10 to 3. The results showed a decrease in the rate of polymerization as the number of side chains in the graft increased. This was followed by a sharp reduction in the number of particles produced. This clearly shows the importance of spacing of the side chains to ensure anchoring of the graft copolymer to the particle surface, which is stronger with a graft containing a smaller number of side chains. If the number of side chains for the PEO with an M_w of 2000 is increased from 3 to 9 (93 wt% of PEO), the rate of polymerization and number of particles decrease. Using a PEO chain with an M_w of 5000 (92 wt% PEO) and three chains per graft gives the same result as the PEO of $M_w = 2000$ with three side chains. Any increase in the number of side chains in the graft results in a reduction in the rate of polymerization and the number of latex particles produced. This clearly shows the importance of spacing of the side chains of the graft copolymer.

Similar results were obtained using a graft copolymer of poly(methyl methacrylate-co-2-hydroxypropyl methacrylate)-graft-poly(ethylene oxide), PMMA(PEO)$_n$, for emulsion polymerization of methyl methacrylate. As with PMSt(PEO)$_n$ graft, the backbone molecular weight had little effect on the rate of polymerization or the number of particles used. The molecular weight of the PEO side chains was varied at constant M_w of the backbone (10 000). Three PEO grafts with M_w of 750, 2000 and 5000 were used. Although the rates of polymerization were similar for the three graft copolymers, the number of particles was significantly lower with the graft containing PEO of $M_w = 750$. This shows that this short PEO chain is not sufficient for stabilization of the particles. The overall content of PEO in the graft has also a large effect. Using the same backbone chain length while changing the weight-percentage of PEO of $M_w = 2000$, it was found that the molecule containing 67 wt% PEO is not sufficient for stabilization of the particles when compared with a graft containing 82 wt% PEO. This shows that a high concentration of PEO in the adsorbed layer is required for effective steric stabilization (see Chapter 4).

The chemical nature of the monomer also plays an important role. For example, stable latexes could be produced using PMSt(PEO)$_n$ graft but not with PMMA(PEO)$_n$ graft.

A novel graft copolymer of hydrophobically modified inulin (INUTEC® SP1) has been used in emulsion polymerization of styrene, methyl methacrylate, butyl acrylate and several other monomers [8]. All latexes were prepared by emulsion polymerization using potassium persulfate as initiator. The z-average particle size was determined by PCS and electron micrographs were also taken.

Emulsion polymerization of styrene or methyl methacrylate showed an optimum weight ratio of INUTEC SP1 to monomer of 0.0033 for PS and 0.001 for PMMA particles. The initiator to monomer ratio was kept constant at 0.00125. The monomer conversion was higher than 85% in all cases. Latex dispersions of PS reaching 50% and of PMMA reaching 40% could be obtained using such a low concentration of INUTEC SP1. Figure 2.1 shows the variation of particle diameter with monomer concentration.

The stability of the latexes was determined by determining the critical coagulation concentration (CCC) using $CaCl_2$. The CCC was low (0.0175–0.05 mol dm^{-3}),

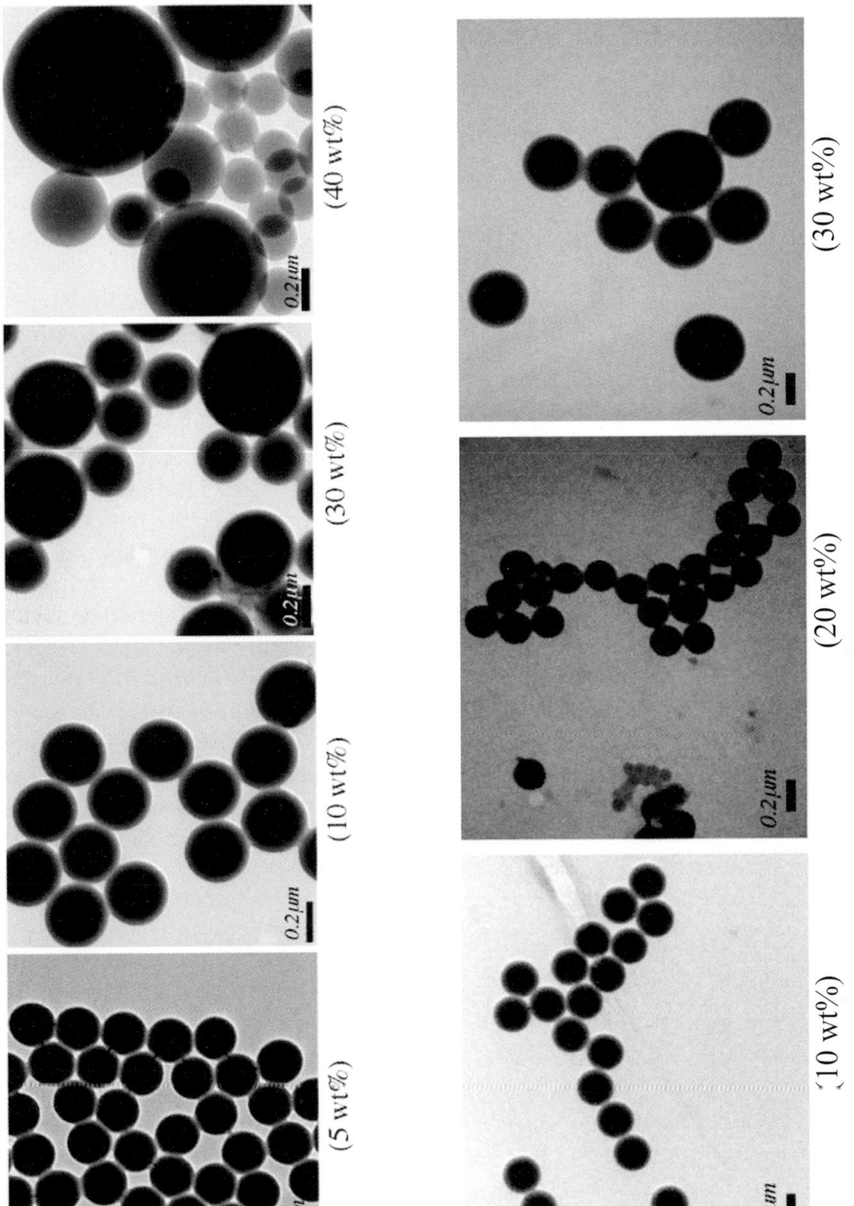

Figure 2.1 Electron micrographs of (a) PS and (b) PMMA latexes.

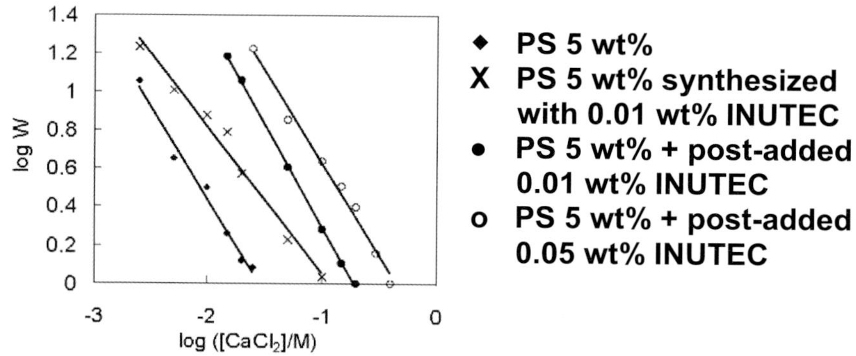

Figure 2.2 Influence of post-addition of INUTEC SP1 on the latex stability.

but it was higher than that for the latex prepared without surfactant. Post-addition of INUTEC SP1 resulted in a large increase in the CCC, as illustrated in Figure 2.2, which shows log W – log C curves (where W is the ratio between the fast flocculation rate constant to the slow flocculation rate constant, referred to as the stability ratio) at various additions of INUTEC SP1.

As with the emulsions, the high stability of the latex when using INUTEC SP1 is due to the strong adsorption of the polymeric surfactant on the latex particles and the formation of strongly hydrated loops and tails of polyfructose that provide effective steric stabilization. Evidence for the strong repulsion produced when using INUTEC SP1 was obtained from atomic force microscopy investigations [9], whereby the force between hydrophobic glass spheres and hydrophobic glass plate, both containing an adsorbed layer of INUTEC SP1, was measured as a function of the separation distance both in water and in the presence of various Na_2SO_4 concentrations. The results are shown in Figures 2.3 and 2.4.

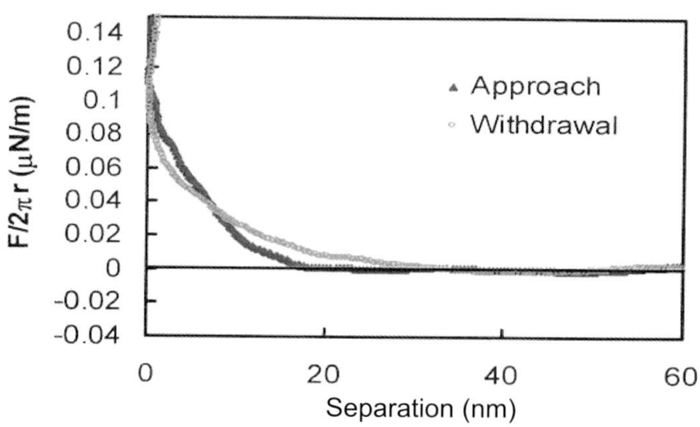

Figure 2.3 Force–distance curves between hydrophobized glass surfaces containing adsorbed INUTEC SP1 in water.

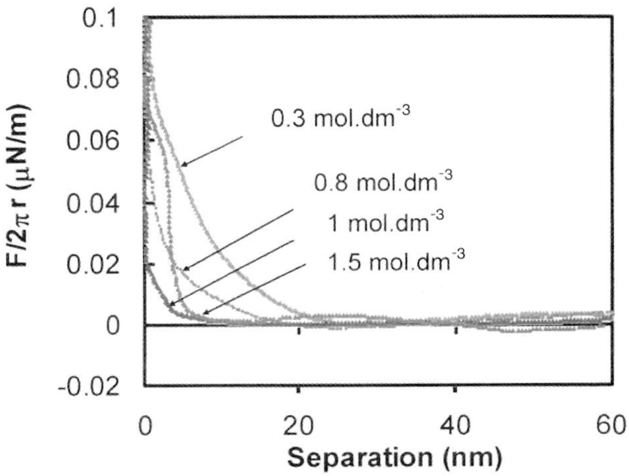

Figure 2.4 Force–distance curves for hydrophobized glass surfaces containing adsorbed INUTEC SP1 at various Na_2SO_4 concentrations.

2.2
Polymeric Surfactants for Stabilization of Preformed Latex Dispersions

For this purpose, polystyrene (PS) latexes were prepared using surfactant-free emulsion polymerization. Two latexes with z-average diameters of 427 and 867 nm (as measured using PCS) that were reasonably monodisperse were prepared [10]. Two polymeric surfactants, Hypermer CG-6 and Atlox 4913 (Uniqema, UK), were used. Both are graft ('comb')-type polymers consisting of a poly(methyl methacrylate)–poly(methacrylic acid) (PMMA–PMA) backbone with methoxy-capped poly(ethylene oxide) (PEO) side chains ($M = 750$ Da). Hypermer CG-6 is the same graft copolymer as Atlox 4913 but it contains a higher proportion of methacrylic acid in the backbone. The average molecular weight of the polymer is ~5000 Da. Figure 2.5 shows a typical adsorption isotherm of Atlox 4913 on the two latexes. Similar results were obtained for Hypermer CG-6 except that the plateau adsorption was lower (1.2 mg m^{-2} compared with 1.5 mg m^{-2} for Atlox 4913). It is likely that the backbone of Hypermer CG-6, which contains more PMA, is more polar and hence less strongly adsorbed. The amount of adsorption was independent of particle size.

The influence of temperature on adsorption is shown in Figure 2.6. The amount of adsorption increases with increase in temperature. This is due to the poorer solvency of the medium for the PEO chains. The PEO chains become less hydrated at higher temperature and the reduction in solubility of the polymer enhances adsorption.

The adsorbed layer thickness of the graft copolymer on the latexes was determined using rheological measurements. Steady-state (shear stress σ–shear rate γ) measurements were carried out and the results were fitted to the Bingham

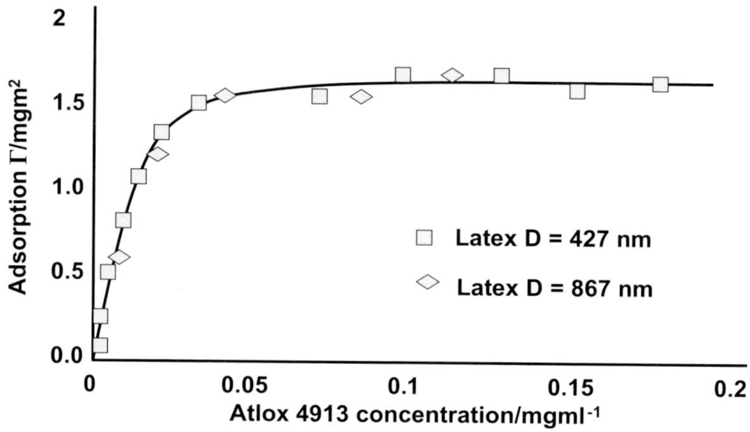

Figure 2.5 Adsorption isotherms of Atlox 4913 on the two latexes at 25 °C.

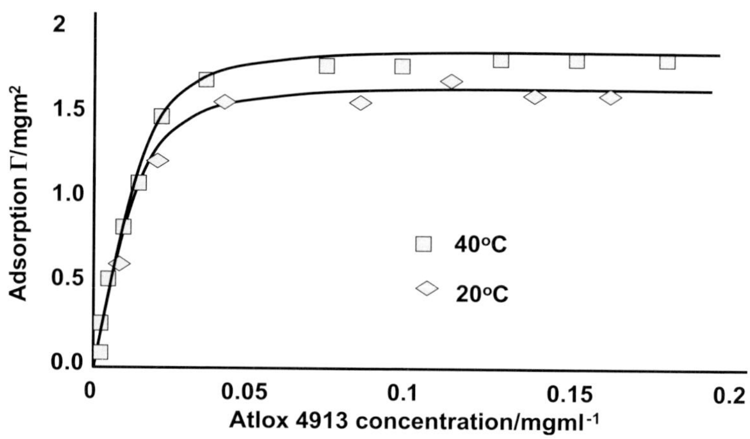

Figure 2.6 Effect of temperature on adsorption of Atlox 4913 on PS.

equation to obtain the yield value σ_β and the high shear viscosity η of the suspension:

$$\sigma = \sigma_\beta + \eta\dot{\gamma} \tag{3}$$

As an illustration, Figure 2.7 shows a plot of σ_β versus volume fraction ϕ of the latex for Atlox 4913. Similar results were obtained for latexes stabilized using Hypermer CG-6.

At any given volume fraction, the smaller latex has a higher σ_β than the larger latex. This is due to the higher ratio of adsorbed layer thickness to particle radius, Δ/R, for the smaller latex. The effective volume fraction of the latex, ϕ_{eff}, is related to the core volume fraction, ϕ, by the equation

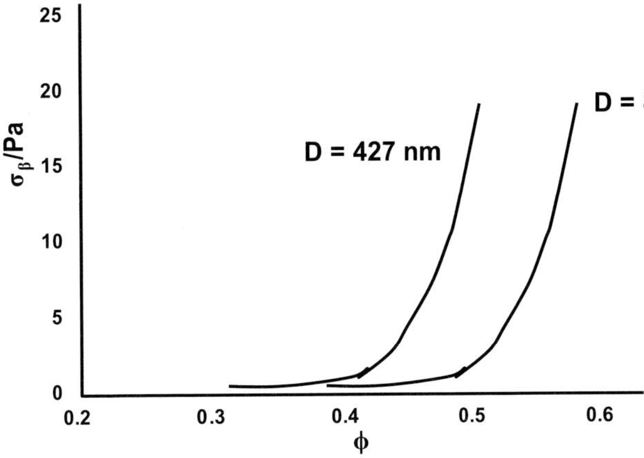

Figure 2.7 Variation of yield stress with latex volume fraction for Atlox 4913.

$$\phi_{\text{eff}} = \phi\left(1 + \frac{\Delta}{R}\right)^3 \tag{4}$$

As discussed earlier, ϕ_{eff} can be calculated from the relative viscosity η_r using the Dougherty–Krieger equation:

$$\eta_r = \left[1 - \left(\frac{\phi_{\text{eff}}}{\phi_p}\right)\right]^{-[\eta]\phi_p} \tag{5}$$

where ϕ_p is the maximum packing fraction, which can be calculated using the following empirical equation [10]:

$$\frac{\eta_r^{\frac{1}{2}} - 1}{\phi} = \left(\frac{1}{\phi_p}\right)(\eta^{\frac{1}{2}} - 1) + 1.25 \tag{6}$$

The results showed a gradual decrease in adsorbed layer thickness Δ with increase in the volume fraction ϕ. For the latex with a diameter D of 867 nm and Atlox 4913, Δ decreased from 17.5 nm at $\phi = 0.36$ to 6.5 nm at $\phi = 0.57$. For Hypermer CG-6 with the same latex, Δ decreased from 11.8 nm at $\phi = 0.49$ to 6.5 nm at $\phi = 0.57$. The decrease in Δ with increase in ϕ may be due to overlap and/or compression of the adsorbed layers as the particles come close to each other at higher volume fraction of the latex.

The stability of the latexes was determined using viscoelastic measurements (see Chapter 6). For this purpose, dynamic (oscillatory) measurements were used to obtain the storage modulus G^*, the elastic modulus G' and the viscous modulus G'' as a function of strain amplitude γ_0 and frequency ω (rad s^{-1}). The method relies on the application of a sinusoidal strain or stress and the resulting stress or strain is measured simultaneously. For a viscoelastic system, the strain and stress sine waves oscillate with the same frequency but out of phase. From the time shift Δt and ω one can obtain the phase angle shift δ.

The ratio of the maximum stress σ_0 to the maximum strain γ_0 gives the complex modulus $|G^*|$:

$$|G^*| = \frac{\sigma_0}{\gamma_0} \tag{7}$$

$|G^*|$ can be resolved into two components: the storage (elastic) modulus G', the real component of the complex modulus, and the loss (viscous) modulus G'', the imaginary component of the complex modulus. The complex modulus can be resolved into G' and G'' using vector analysis and the phase angle shift δ:

$$G' = |G^*| \cos \delta \tag{8}$$

$$G'' = |G^*| \sin \delta \tag{9}$$

G' is measured as a function of electrolyte concentration and/or temperature to assess the latex stability. As an illustration, Figure 2.8 shows the variation of G' with temperature for latex stabilized with Atlox 4913 in the absence of any added electrolyte and in the presence of 0.1, 0.2 and 0.3 mol dm^{-3} Na_2SO_4.

In the absence of electrolyte, G' showed no change with temperature up to 65 °C in the presence of 0.1 mol dm^{-3} Na_2SO_4, G' remained constant up to 40 °C, above which it increased with further increase in temperature. This temperature is denoted the critical flocculation temperature (CFT). The CFT decreases with increase in electrolyte concentration, reaching ~30 °C in 0.2 and 0.3 mol dm^{-3} Na_2SO_4. This reduction in CFT with increase in electrolyte concentration is due to the reduction in solvency of the PEO chains with increase in electrolyte concentrations. The latex stabilized with Hypermer CG-6 gave relatively higher CFT values compared with that stabilized using Atlox 4913.

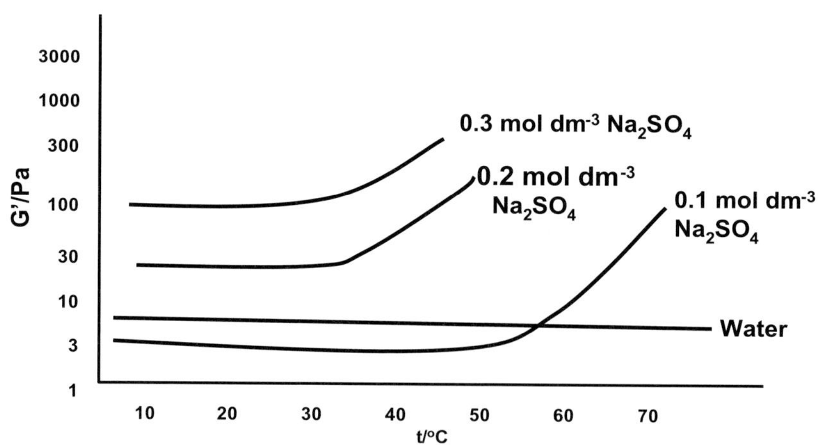

Figure 2.8 Variation of G' with temperature in water and at various Na_2SO_4 concentrations.

2.3 Dispersion Polymerization

This method is usually applied for the preparation of non-aqueous latex dispersions and hence it is referred to as NAD. The method has also been adapted to prepare aqueous latex dispersions by using an alcohol–water mixture.

In the NAD process, the monomer, normally an acrylic, is dissolved in a non-aqueous solvent, normally an aliphatic hydrocarbon, and an oil-soluble initiator and a stabilizer (to protect the resulting particles from flocculation, sometimes referred to as 'protective colloid') is added to the reaction mixture. The most successful stabilizers used in NAD are block and graft copolymers. These block and graft copolymers are assembled in a variety of ways to provide the molecule with an 'anchor chain' and a stabilizing chain. The anchor chain should be sufficiently insoluble in the medium and has a strong affinity to the polymer particles produced. In contrast, the stabilizing chain should be soluble in the medium and strongly solvated by its molecules to provide effective steric stabilization (see Chapter 4). The length of the anchor and stabilizing chains has to be carefully adjusted to ensure strong adsorption (by multipoint attachment of the anchor chain to the particle surface) and a sufficiently 'thick' layer of the stabilizing chain that prevents close approach of the particles to a distance where the van der Waals attraction becomes strong. The criteria for effective steric stabilization by block and graft copolymers are discussed in detail in Chapter 4. Several configurations of block and graft copolymers are possible, as illustrated in Figure 2.9.

Typical preformed graft stabilizers based on poly(12-hydroxystearic acid) (PHS) are simple to prepare and effective in NAD polymerization. Commercial 12-hydroxystearic acid contains 8–15% of palmitic and stearic acids, which limits the molecular weight during polymerization to an average of 1500–2000. This oligomer may be converted to a 'macromonomer' by reacting the carboxylic group with glycidyl methacrylate. The macromonomer is then copolymerized with an equal weight of methyl methacrylate (MMA) or similar monomer to give a 'comb' graft copolymer with an average molecular weight of 10 000–20 000. The graft copolymer contains on average 5–10 PHS chains pendent from a polymeric anchor backbone of PMMA. This graft copolymer can stabilize latex particles of various monomers. The major limitation of the monomer composition is that the polymer produced should be insoluble in the medium used.

Several other examples of block and graft copolymers that are used in dispersion polymerization are given in Table 2.4, which also shows the continuous phase and disperse polymers that can be used with these polymers.

Two main criteria must be considered in the process of dispersion polymerization: (i) the insolubility of the polymer formed in the continuous phase and (ii) the solubility of the monomer and initiator in the continuous phase. Initially, dispersion polymerization starts as a homogeneous system but, after sufficient polymerization, the insolubility of the resulting polymer in the medium forces them to precipitate. Initially polymer nuclei are produced, which then grow into polymer particles. The latter are stabilized against aggregation by the block or graft copolymer that is added to the continuous phase before the process of

Table 2.4 Examples of block and graft copolymers.

Polymeric surfactant	Continuous phase	Disperse polymer
Polystyrene-block-polydimethylsiloxane	Hexane	Polystyrene
Polystyrene-block-poly(methacrylic acid)	Ethanol	Polystyrene
Polybutadiene-graft-poly(methacrylic acid)	Ethanol	Polystyrene
Poly(2-ethylhexyl acrylate)-graft-poly(vinyl acetate)	Aliphatic hydrocarbon	Poly(methyl methacrylate)
Polystyrene-block-poly(*tert*-butylstyrene)	Aliphatic hydrocarbon	Polystyrene

polymerization starts. It is essential to choose the right block or graft copolymer, which should have a strong anchor chain A and a good stabilizing chain B, as shown schematically in Figure 2.9.

Dispersion polymerization may be considered as a heterogeneous process which may include emulsion, suspension, precipitation and dispersion polymerization. In dispersion and precipitation polymerization, the initiator must be soluble in the continuous phase, whereas in emulsion and suspension polymerization, the initiator is chosen to be soluble in the disperse phase of the monomer. A comparison of the rate of polymerization of methyl methacrylate at 80 °C for the three systems was given by Barrett and Thomas [11], as illustrated in Figure 2.10 The rate of dispersion polymerization is much faster than that of

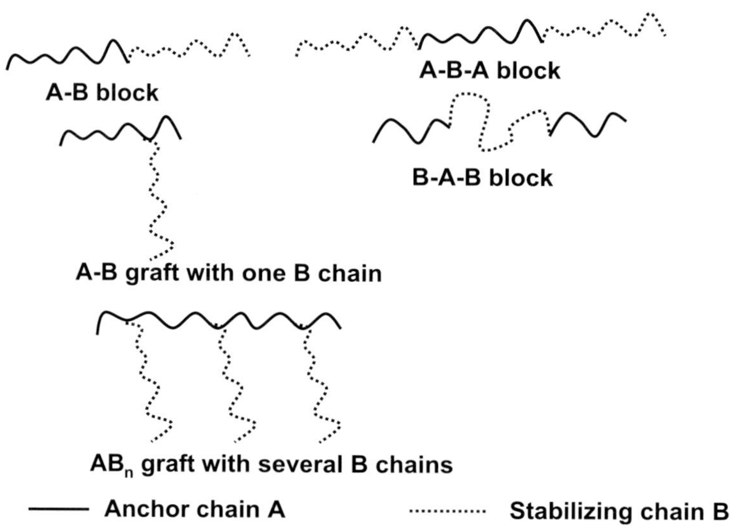

Figure 2.9 Configurations of block and graft copolymers.

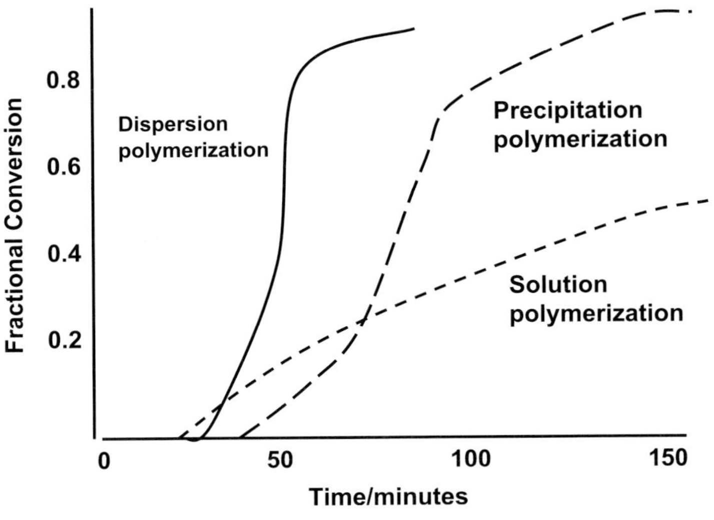

Figure 2.10 Comparison of rates of polymerization.

precipitation or solution polymerization. The enhancement of the rate in precipitation polymerization over solution polymerization has been attributed to the hindered termination of the growing polymer radicals.

2.3.1 Mechanism of Dispersion Polymerization

Several mechanisms have been proposed to explain the mechanism of emulsion polymerization, but no single mechanism can explain all occurrences. Barrett and Thomas [11] suggested that particles are formed in emulsion polymerization by two main steps: (i) initiation of monomer in the continuous phase and subsequent growth of the polymer chains until the latter become insoluble; this process clearly depends on the nature of the polymer and medium; and (ii) the growing oligomeric chains associate with each other, forming aggregates which below a certain size are unstable and become stabilized by the block or graft copolymer added. As mentioned earlier, this aggregative nucleation theory cannot explain all occurrences in dispersion polymerization. An alternative mechanism based on Napper and co-workers' theory [4, 5] for aqueous emulsion polymerization can be adapted to the process of dispersion polymerization. This theory includes coagulation of the nuclei formed and not just association of the oligomeric species. The precursor particles (nuclei), being unstable, can undergo one of the following events to become colloidally stable: (i) homocoagulation, i.e. collision with other precursor particles; (ii) growth by propagation, adsorption of stabilizer; and (iii) swelling with monomer. The nucleation-terminating events are diffusional capture of oligomers and heterocoagulation.

The number of particles formed in the final latex does not depend on particle nucleation alone, since other steps are involved which determine how many precursor particles created are involved in the formation of a colloidally stable particle. This clearly depends on the effectiveness of the block or graft copolymer used in stabilizing the particles (see below).

2.3.2
Influence of Polymeric Surfactant Concentration and Molecular Weight on Particle Formation

In most cases, an increase in polymeric surfactant concentration (at any given monomer amount) results in the production of larger number of particles with smaller size. This is to be expected since the larger number of particles with smaller size (i.e. larger total surface area of the disperse particles) require more polymeric surfactant for their formation. The molecular weight of the polymeric surfactant can also influence the number of particles formed. For example Dawkins and Taylor [12] found that in dispersion polymerization of styrene in hexane, increasing the molecular weight of the block copolymer of polydimethylsiloxane-block-polystyrene resulted in the formation of smaller particles, which was attributed to the more effective steric stabilization by the higher molecular weight block.

2.3.3
Effect of Monomer Solubility and Concentration in the Continuous Phase

A systematic study of the effect of monomer solubility and concentration in the continuous phase was carried out by Antl *et al.* [13]. Dispersion polymerization of methyl methacrylate in hexane mixed with a high boiling point aliphatic hydrocarbon was investigated using poly(12-hydroxystearic acid)–glycidyl methacrylate block copolymer. They found that the methyl methacrylate concentration had a drastic effect on the size of the particles produced. When the monomer concentration was kept below 8.5%, very small particles (80 nm) were produced and these remained very stable. However, between 8.5 and 35% monomer, the latex produced was initially stable but flocculated during polymerization. An increase in monomer concentration from 35 to 50% resulted in the formation of a stable latex, but the particle size increased sharply from 180 nm to 2.6 μm as the monomer concentration increased. The authors suggested that the final particle size and stability of the latex are strongly affected by increased monomer concentration in the continuous phase. The presence of monomer in the continuous phase increases the solvency of the medium for the polymer formed. In a good solvent for the polymer, the growing chain is capable of reaching higher molecular weight before it is forced to phase separate and precipitate.

NAD polymerization is carried in two steps: (i) seed stage: the diluent, portion of the monomer, portion of dispersant and initiator (azo or peroxy type) are heated to form an initial low-concentration fine dispersion; and (ii) growth stage: the remaining monomer together with more dispersant and initiator are then fed over the course of several hours to complete the growth of the particles. A small amount of transfer agent is usually added to control the molecular weight. Excellent control of particle size is achieved by proper choice of the designed dispersant and correct distribution of dispersant between the seed and growth stages. NAD acrylic polymers are applied in automotive thermosetting polymers and hydroxy monomers may be included in the monomer blend used.

2.3.4 Stability/Instability of the Resulting Latex

Two main factors must be considered when considering the long-term stability of a non-aqueous polymer dispersion. The first and very important factor is the nature of the 'anchor chain' A. As mentioned above, this should have a strong affinity to the latex produced and in most cases it can be designed to be 'chemically' attached to the polymer surface. Once this criterion is satisfied, the second and also important factor in determining the stability is the solvency of the medium for the stabilizing chain B. As will be discussed in detail, the solvency of the medium is characterized by the Flory–Huggins interaction parameter χ. Three main conditions can be identified: $\chi < 0.5$ (good solvent for the stabilizing chain); $\chi > 0.5$ (poor solvent for the stabilizing chain); and $\chi = 0.5$ (referred to as the θ-solvent). Clearly, to maintain stability of the latex dispersion, the solvent must be better than a θ-solvent. The solvency of the medium for the B chain is affected by addition of a non-solvent and/or temperature changes. It is essential, therefore, to determine the critical volume fraction (CFV) of a non-solvent above which flocculation (sometimes referred to as incipient flocculation) occurs. One should also determine the critical flocculation temperature at any given solvent composition, below which flocculation occurs. The correlation between CFV or CFT and the flocculation of the non-aqueous polymer dispersion has been demonstrated by Napper [14], who investigated the flocculation of poly(methyl methacrylate) dispersions stabilized by poly(12-hydroxystearic acid) or poly(*n*-lauryl methacrylate-co-glycidyl methacrylate) in hexane by adding a non-solvent such as ethanol or propanol and cooling the dispersion. The dispersions remained stable until the addition of ethanol transformed the medium to a θ-solvent for the stabilizing chains in solution. However, flocculation did occur under conditions of slightly better than θ-solvent for the chains. The same was found for the CFT, which was 5–15 K above the θ-temperature. This difference was accounted for by the polydispersity of the polymer chains. The θ-condition is usually determined by cloud point measurements and the least soluble component will precipitate first, giving values that are lower than the CFV or higher than the CFT.

2.3.5 Particle Formation in Polar Media

The process of dispersion polymerization has been applied in many cases using completely polar solvents such as alcohols or alcohol–water mixtures [15]. The results obtained showed completely different behavior when compared with dispersion polymerization in non-polar media. For example, results obtained by Lock using styrene as monomer and hydroxypropylcellulose as stabilizer showed a linear increase in particle diameter with increase in the weight-percentage of the monomer. There was no region of monomer concentration where instability occurred (as has been observed for the dispersion polymerization of methyl methacrylate in aliphatic hydrocarbons). Replacing water in the continuous phase with 2-methoxyethanol, Lok and Oble were able to grow large, monodisperse particles up to 15 μm in diameter. They concluded from these results that the polarity of the medium is the controlling factor in the formation of particles and their final size. The authors suggested a mechanism in which the polymeric surfactant molecule grafts to the polystyrene chain, forming a physically anchored stabilizer (nuclei). These nuclei grow to form the polymer particles. Paine [16] carried out dispersion polymerization of styrene by systematically increasing the alcohol chain length from methanol to octadecanol and using hydroxypropylcellulose as stabilizer. The results showed an increase in particle diameter with increase in number of carbon atoms in the alcohol, reaching a maximum when hexanol was used as the medium, after which there was a sharp decrease in the particle diameter with further increase in the number of carbon atoms in the alcohol. Paine explained his results in terms of the solubility parameter of the dispersion medium. The largest particles are produced when the solubility parameter of the medium is closest to those of styrene and hydroxpropylcellulose.

References

1 D.C. Blakely, *Emulsion Polymerization*, Elsevier Applied Science, London (1975).
2 K.E.J. Barrett (ed.), *Dispersion Polymerization in Organic Media*, John Wiley & Sons, Ltd, Chichester (1975).
3 W.V. Smith and R.H. Ewart, *J. Chem. Phys.*, *16*, 592 (1948).
4 G. Litchi, R.G. Gilbert and D.H. Napper, *J. Polym. Sci.*, *21*, 269 (1983).
5 P.J. Feeney, D.H. Napper and R.G. Gilbert, *Macromolecules*, *17*, 2520 (1984); *20*, 2922 (1987).
6 I. Piirma, *Polymeric Surfactants*, Marcel Dekker, New York (1992).
7 I. Pirma and J.R. Lenzotti, *Br. Polym. J.*, *21*, 45 (1989).
8 J. Nestor, J. Esquena, C. Solans, B. Levecke, K. Booten and Th.F. Tadros, *Langmuir*, *21*, 4837 (2005).
9 J. Nestor, J. Esquena, C. Solans, P.F. Luckham, B. Levecke and Th.F. Tadros, *J. Colloid Interface Sci.*, *311*, 430 (2007).
10 W. Liang, G. Bognolo and Th.F. Tadros, *Langmuir*, *11*, 2899 (1995).
11 K.E.J. Barrett and H.R. Thomas, *J. Polym. Sci., Part A1*, *7*, 2627 (1969).
12 J.V. Dawkins and G. Taylor, *Polymer*, *20*, 173 (1987).
13 L. Antl, J.W. Goodwin, R.D. Hill, R.H. Ottewill, S.M. Owen, S. Papworth and J.A. Waters, *Colloids Surf.*, *1*, 67 (1986).
14 D.H. Napper, *Polmeric Stabilization of Colloidal Dispersions*, Academic Press, London (1983).
15 K.P. Lok and C.K. Ober, *Can. J. Chem.*, *63*, 209 (1985).
16 A.J. Paine, *J. Polym. Sci., Part A*, *28*, 2485 (1990).

3
Pigment Dispersion

As mentioned in Chapter 1, a paint may be considered as a colloidal dispersion of a pigment (the disperse phase) in a medium that may be aqueous or non-aqueous that contains a polymer (film former) and/or latex particles. The state of dispersion of a pigment in a paint is vital as it determines its optical properties (e.g. color), flow properties (rheology), durability, opacity, gloss and storage stability. This chapter deals with the dispersion of a powder in a liquid.

Dispersion methods are used for the preparation of suspensions of preformed particles: the term dispersion is used to refer to the complete process of incorporating the solid into a liquid such that the final product consists of fine particles distributed throughout the dispersion medium. The role of surfactants (or polymers) in the dispersion can be seen from consideration of the stages involved [1, 2]. Three stages have been considered [3]: wetting of the powder by the liquid, breaking of the aggregates and agglomerates and comminution (milling) of the resulting particles into smaller units. These three stages are considered below.

3.1
Powder Wetting

Before describing the wetting of powders (which consist of aggregates, where the particles are attached by their 'faces', or agglomerates, where the particles are attached by their 'corners'), it is essential to describe the fundamental process of wetting in general and also other factors such as adhesion of liquids to surfaces, spreading of liquids on substrates and the importance of the surface energy of the solid. These fundamental principles are briefly described below.

3.1.1
Wetting of Substrates

Wetting is a fundamental process in which one fluid phase is displaced completely or partially by another fluid phase from the surface of a solid. A useful parameter to describe wetting is the contact angle θ of a liquid drop on a solid substrate. If the liquid makes no contact with the solid, i.e. $\theta = 180^\circ$, the solid is

Colloids and Interface Science Series, Vol. 6
Colloids in Paints. Tharwat F. Tadros

ISBN: 978-3-527-31466-9

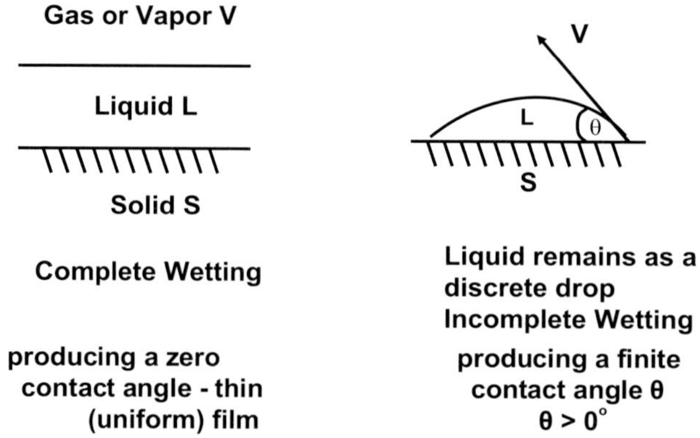

Figure 3.1 Illustration of complete and partial wetting.

referred to as non-wettable by the liquid in question. This may be the case for a perfectly hydrophobic surface with a polar liquid such as water. However, when $180^\circ > \theta > 90^\circ$, one may refer to a case of poor wetting. When $0^\circ < \theta < 90^\circ$, partial (incomplete) wetting is the case, whereas when $\theta = 0^\circ$ complete wetting occurs and the liquid spreads on the solid substrate forming a uniform liquid film. The cases of partial and complete wetting are shown schematically in Figure 3.1, for a liquid on a perfectly smooth solid substrate.

The utility of contact angle measurements depends on equilibrium thermodynamic arguments (static measurements) using the well-known Young's equation [4]. The value depends on (i) the history of the system and (ii) whether the liquid is tending to advance across or recede from the solid surface (advancing angle θ_A, receding angle θ_R; usually $\theta_A > \theta_R$).

Under equilibrium, the liquid drop takes the shape that minimizes the free energy of the system. Three interfacial tensions can be identified: γ_{SV}, solid/vapor area A_{SV}; γ_{SL}, solid/liquid area A_{SL}; γ_{LV}, liquid/vapor area A_{LV}. A schematic representation of the balance of tensions at the solid/liquid/vapor interface is shown in Figure 3.2. The contact angle is that formed between the planes tangent to the

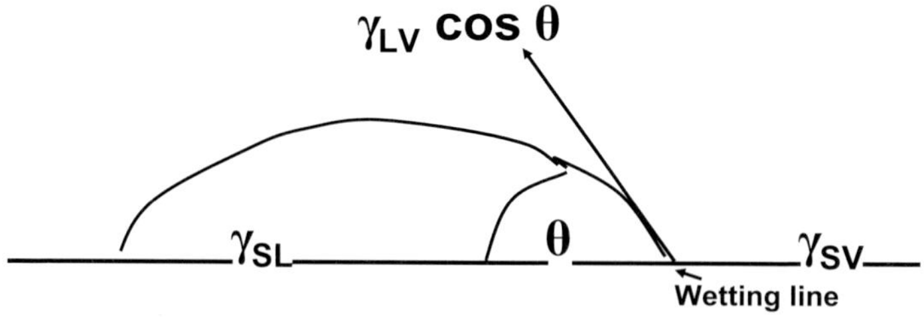

Figure 3.2 Schematic representation of contact angle and wetting line.

surfaces of the solid and liquid at the wetting perimeter. Here, solid and liquid are simultaneously in contact with each other and the surrounding phase (air or vapor of the liquid). The wetting perimeter is referred to as the three-phase line or wetting line. In this region, there is an equilibrium between vapor, liquid and solid.

$\gamma_{SV}A_{SV} + \gamma_{SL}A_{SL} + \gamma_{LV}A_{LV}$ should be a minimum at equilibrium and this leads to the well-known Young's equation:

$$\gamma_{SV} = \gamma_{SL} + \gamma_{LV} \cos \theta \quad (1)$$

$$\cos \theta = \frac{\gamma_{SV} - \gamma_{SL}}{\gamma_{LV}} \quad (2)$$

The contact angle θ depends on the balance between the solid/vapor (γ_{SV}) and solid/liquid (γ_{SL}) interfacial tensions. The angle which a drop assumes on a solid surface is the result of the balance between the adhesion force between solid and liquid and the cohesive force in the liquid:

$$\gamma_{LV} \cos \theta = \gamma_{SV} - \gamma_{SL} \quad (3)$$

If there is no interaction between solid and liquid:

$$\gamma_{SL} = \gamma_{SV} + \gamma_{LV} \quad (4)$$

i.e. $\cos \theta = -1$ or $\theta = 180°$. If there is strong interaction between solid and liquid (maximum wetting), the latter spreads until Young's equation is satisfied:

$$\gamma_{LV} = \gamma_{SV} - \gamma_{SL} \quad (5)$$

i.e. $\cos \theta = 1$ or $\theta = 0°$; the liquid is described to spread spontaneously on the solid surface.

When the surface of the solid is in equilibrium with the liquid vapor, we can consider the spreading pressure π, as is illustrated in Figure 3.3.

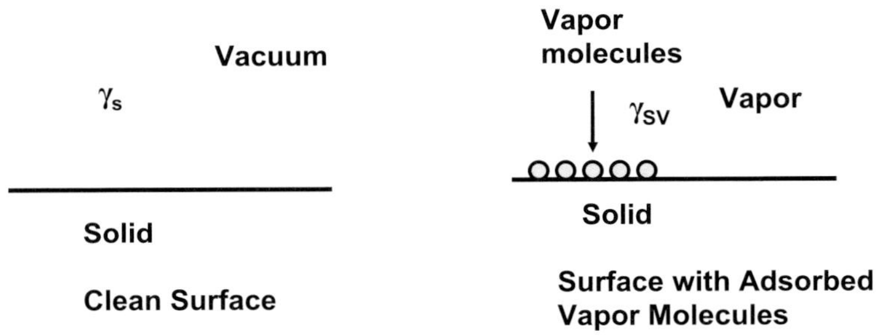

Figure 3.3 Schematic representation of the spreading pressure.

The solid surface tension is lowered as a result of adsorption of vapor molecules:

$$\pi = \gamma_s - \gamma_{SV} \tag{6}$$

Young's equation can be written as

$$\gamma_{LV} \cos\theta = \gamma_s - \gamma_{SL} - \pi \tag{7}$$

3.1.2
Adhesion Tension

There is no direct way by which γ_{SV} or γ_{SL} can be measured. The difference between γ_{SV} and γ_{SL} can be obtained from contact angle measurements ($= \gamma_{LV} \cos\theta$). This difference is referred to as wetting tension or adhesion tension:

$$\text{adhesion tension} = \gamma_{SV} - \gamma_{SL} = \gamma_{LV} \cos\theta \tag{8}$$

Gibbs [5] defined the adhesion tension τ as the difference between the surface pressure of the solid/liquid and that between the solid/vapor interface:

$$\tau = \pi_{SL} - \pi_{SV} \tag{9}$$

$$\pi_{SV} = \gamma_s - \gamma_{SV} \tag{10}$$

$$\pi_{SL} = \gamma_s - \gamma_{SL} \tag{11}$$

$$\tau = \gamma_{SV} - \gamma_{SL} = \gamma_{LV} \cos\theta \tag{12}$$

3.1.3
Work of Adhesion, W_a

The work of adhesion is a direct measure of the free energy of interaction between solid and liquid [6, 7]; this is illustrated in Figure 3.4.

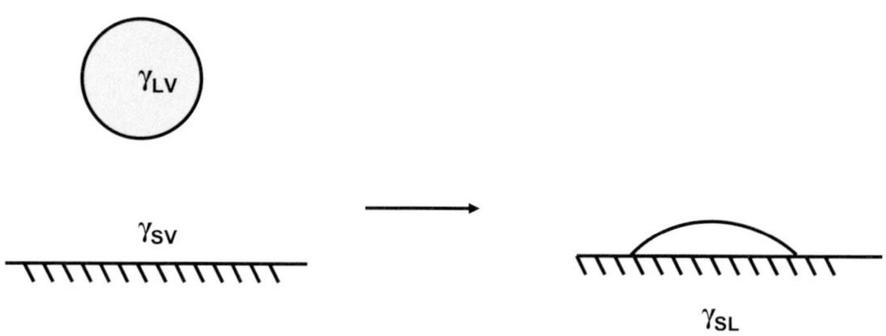

Figure 3.4 Representation of adhesion of a drop on a solid substrate.

$$W_a = (\gamma_{LV} + \gamma_{SV}) - \gamma_{SL} \quad (13)$$

Using Young's equation:

$$W_a = \gamma_{LV} + \gamma_{SV} - \gamma_{LV} \cos \theta = \gamma_{LV}(\cos \theta + 1) \quad (14)$$

The work of adhesion depends on γ_{LV}, the liquid/vapor surface tension, and θ, the contact angle between liquid and solid.

3.1.4
The Work of Cohesion

The work of cohesion, W_c, is the work of adhesion when the two surfaces are the same. Consider a liquid cylinder with unit cross-sectional area as represented in Figure 3.5.

$$W_c = 2\gamma_{LV} \quad (15)$$

For adhesion of a liquid on a solid, $W_a \approx W_c$ or $\theta = 0°$ ($\cos \theta = 1$).

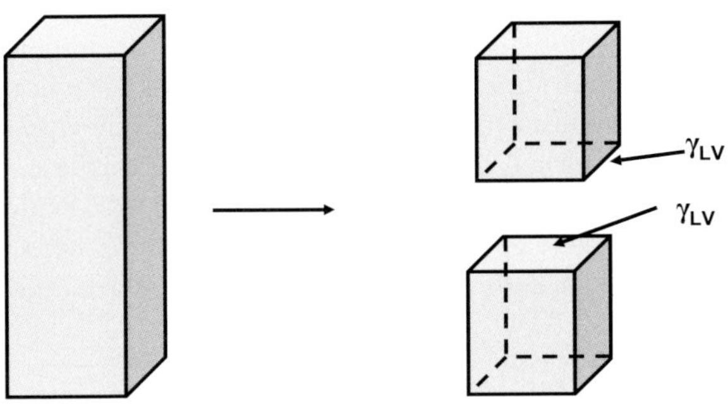

Figure 3.5 Schematic representation of subdivision of a liquid cylinder.

3.1.5
Spreading Coefficient, *S*

Harkins [8, 9] defined the spreading coefficient as the work required to destroy unit area of SL and LV and leaves unit area of bare solid SV, as illustrated in Figure 3.6: spreading coefficient S = surface energy of final state – surface energy of initial state:

$$S = \gamma_{SV} - (\gamma_{SL} + \gamma_{LV}) \quad (16)$$

Using Young's equation:

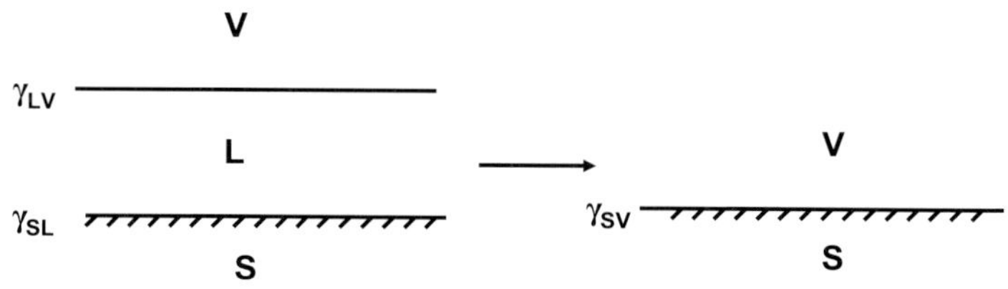

Figure 3.6 Schematic representation of the spreading coefficient *S*.

$$\gamma_{SV} = \gamma_{SL} + \gamma_{LV} \cos \theta \tag{17}$$

$$S = \gamma_{LV}(\cos \theta - 1) \tag{18}$$

If S is zero (or positive), i.e. $\theta = 0$, the liquid will spread until it completely wets the solid. If S is negative, i.e. $\theta > 0$, only partial wetting occurs. Alternatively, one can use the equilibrium (final) spreading coefficient. For dispersion of powders into liquids, one usually requires complete spreading, i.e. θ should be zero.

3.1.6
Contact Angle Hysteresis

For a liquid spreading on a uniform, non-deformable solid (idealized case), there is only one contact angle – the equilibrium value. With real systems (practical solids), a number of stable contact angles can be measured. Two relatively reproducible angles can be measured: largest – advancing angle θ_A; and smallest – receding angle θ_R; this is illustrated in Figure 3.7. θ_A is measured by advancing the periphery of a drop over a surface (e.g. by adding more liquid to the drop); θ_R is measured by pulling the liquid back; $\theta_A - \theta_R$ is referred to as contact angle hysteresis.

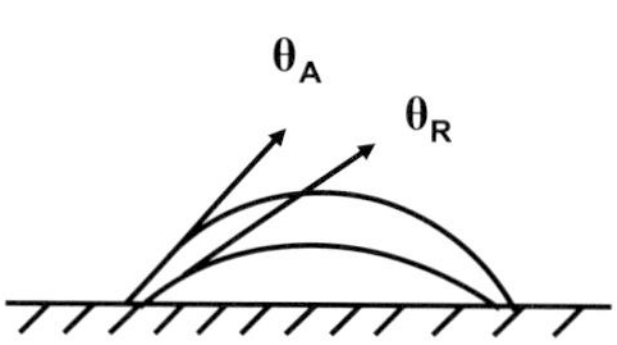

Figure 3.7 Schematic representation of advancing and receding angles.

3.1.7
Reasons for Hysteresis

Reasons for hysteresis are (1) penetration of wetting liquid into pores during advancing contact angle measurements; and (2) surface roughness: the first and rear edges both meet the liquid with some intrinsic angle θ_0 (microscopic contact angle). The macroscopic angles θ_A and θ_R vary significantly. This is best illustrated for a surface-inclined at an angle α from the horizontal (Figure 3.8).

θ_0 values are determined by contact of liquid with the 'rough' valleys (microscopic contact angle); θ_A and θ_R are determined by contact of liquid with arbitrary parts on the surface (peak or valley). Surface roughness can be accounted for by comparing the 'real' area of the surface A with that of the projected (apparent) area A':

$$r = \frac{A}{A'} \tag{19}$$

where A = area of surface taking into account all peaks and valleys and A' = apparent area (same macroscopic dimension); $r > 1$:

$$\cos\theta = r\cos\theta_0 \tag{20}$$

where θ = macroscopic contact angle and θ_0 = microscopic contact angle.

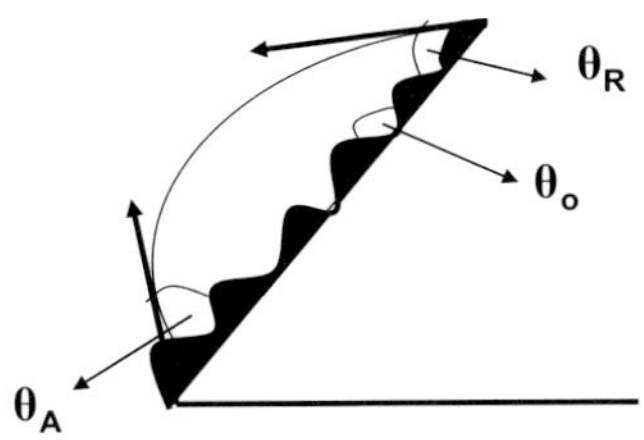

Figure 3.8 Representation of a drop profile on a rough surface.

3.1.8
Wenzel's Equation [10]

Wenzel's equation is

$$\cos\theta = r\left(\frac{\gamma_{SV} - \gamma_{SL}}{\gamma_{LV}}\right) \tag{21}$$

If $\cos\theta$ is negative on a smooth surface ($\theta > 90°$), it becomes more negative on a rough surface (θ is larger) and surface roughness reduces wetting. If $\cos\theta$ is positive on a smooth surface ($\theta < 90°$), it becomes more positive on a rough surface (θ is smaller) and roughness enhances wetting.

3.1.9
Surface Heterogeneity

Most practical surfaces are heterogeneous, consisting of 'islands' or 'patches' with different surface energies. As the drop advances on such a surface, the edge of the drop tends to stop at the boundary of the 'island'. The advancing angle will be associated with the intrinsic angle of the high contact angle region. The receding angle will be associated with the low contact angle region. If the heterogeneities are very small compared with the dimensions of the liquid drop, one can define a composite contact angle using Cassie's equation [11, 12]:

$$\cos\theta = Q_1 \cos\theta_1 + Q_2 \cos\theta_2 \tag{22}$$

where Q_1 = fraction of surface having contact angle θ_1, Q_2 = fraction of surface having contact angle θ_2 and θ_1 and θ_2 are the maximum and minimum possible angles, respectively.

3.1.10
Critical Surface Tension of Wetting

A systematic way of characterizing the 'wettability' of a surface was introduced by Zisman [13]. For a given substrate and for a series of related liquids (e.g. *n*-alkanes, siloxanes and dialkyl ethers), a plot of $\cos\theta$ versus γ_{LV} gives a straight line, as shown schematically in Figure 3.9.

Extrapolation of the straight line to $\cos\theta = 1$ ($\theta = 0$) gives the critical surface tension of wetting, γ_c. Any liquid with $\gamma_{LV} < \gamma_c$ will give $\theta = 0$, i.e. it wets the surface completely; γ_c is the surface tension of a liquid that just spreads on the substrate to give complete wetting.

The above linear relationship can be represented by the following empirical equation:

$$\cos\theta = 1 + b(\gamma_{LV} - \gamma_c) \tag{23}$$

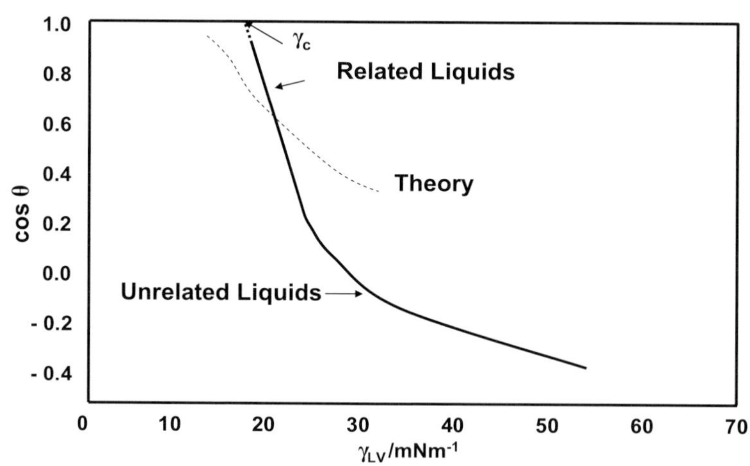

Figure 3.9 Illustration of the wettability of a surface.

High-energy solids, e.g. glass, give high γ_c (>40 mN m^{-1}). Low-energy solids, e.g. hydrophobic surfaces, give lower γ_c (~30 mN m^{-1}). Very low-energy solids such as Teflon (polytetrafluoroethylene, PTFE) give lower γ_c (<25 mN m^{-1}).

The above concept explains the degree of wettability of the above surfaces with water, with glass being the easiest and Teflon being the most difficult to wet.

3.1.11
Effect of Surfactant Adsorption

Surfactants lower the surface tension of water, γ, and they adsorb at the solid/liquid interface. A plot of γ_{LV} versus log C (where C is the surfactant concentration) results in a gradual decrease in γ_{LV} followed by a linear decrease in γ_{LV} with log C (just below the critical micelle concentration, CMC), and when the CMC is reached γ_{LV} remains virtually constant. This is shown schematically in Figure 3.10.

From the slope of the linear portion of the γ–log C curve (just below the CMC), one can obtain the surface excess (number of moles of surfactant per unit area at the Liquid/Air interface). Using the Gibbs adsorption isotherm:

$$\frac{d\gamma}{d \log C} = -2.303RT\Gamma \tag{24}$$

where Γ = surface excess (mol m^{-2}), R = gas constant and T = absolute temperature. From Γ, one can obtain the area per molecule:

$$\text{area per molecule} = \frac{1}{\Gamma N_{av}}(\text{m}^2) = \frac{10^{18}}{\Gamma N_{av}}(\text{nm}^2) \tag{25}$$

Most surfactants produce a vertically oriented monolayer just below the CMC. The area per molecule is usually determined by the cross-sectional area of the

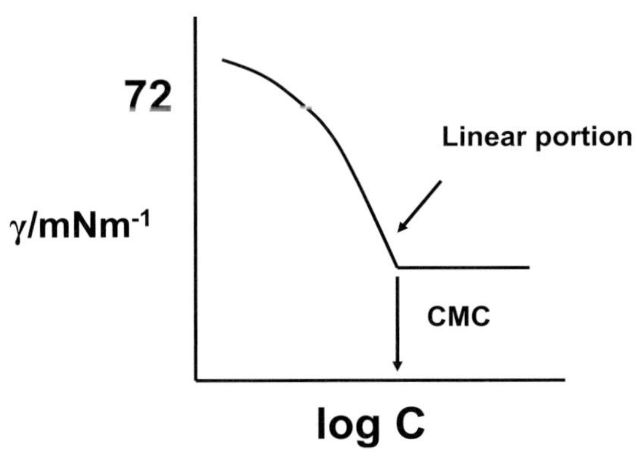

Figure 3.10 Surface tension–log C curves.

head group. For ionic surfactants containing, say, an $-OSO_3^-$ or $-SO_3^-$ head group, the area per molecule is in the region of 0.4 nm^2. For nonionic surfactants containing several moles of ethylene oxide, the area per molecule can be much larger (1–2 nm^2). Surfactants will also adsorb at the solid/liquid interface. For hydrophobic surfaces, the main driving force for adsorption is by hydrophobic bonding. This results in lowering of the contact angle of water on the solid surface. For hydrophilic surfaces, adsorption occurs via the hydrophilic group, e.g. cationic surfactants on silica. Initially, the surface becomes more hydrophobic and the contact angle θ increases with increase in surfactant concentration. However, at higher cationic surfactant concentration, a bilayer is formed by hydrophobic interaction between the alkyl groups and the surface becomes more and more hydrophilic, and eventually the contact angle reaches zero at high surfactant concentrations.

Smolders [14] suggested the following relationship for change of θ with C:

$$\frac{d\gamma_{LV} \cos\theta}{d \ln C} = \frac{d\gamma_{SV}}{d \ln C} - \frac{d\gamma_{SL}}{d \ln C} \qquad (26)$$

Using the Gibbs equation:

$$\sin\theta\left(\frac{d\gamma}{d \ln C}\right) = RT(\Gamma_{SV} - \Gamma_{SL} - \gamma_{LV}\cos\theta) \qquad (27)$$

Since $\gamma_{LV} \sin\theta$ is always positive, then $d\theta/d \ln C$ will always have the same sign as the right-hand side of equation 27). Three cases may be distinguished: (i) $d\theta/d \ln C < 0$, $\Gamma_{SV} < \Gamma_{SL} + \Gamma_{LV}\cos\theta$; addition of surfactant improves wetting; (ii) $d\theta/d \ln C = 0$, $\Gamma_{SV} = \Gamma_{SL} + \Gamma_{LV}\cos\theta$; surfactant has no effect on wetting; (iii) $d\theta/d \ln C > 0$, $\Gamma_{SV} > \Gamma_{SL} + \Gamma_{LV}\cos\theta$; surfactant causes dewetting.

3.1.12 Wetting of Powders by Liquids

Wetting of powders by liquids is very important in their dispersion, e.g. in the preparation of concentrated suspensions. The particles in a dry powder form either aggregates or agglomerates, as illustrated in Figure 3.11.

It is essential in the dispersion process to wet both external and internal surfaces and displace the air entrapped between the particles. Wetting is achieved by

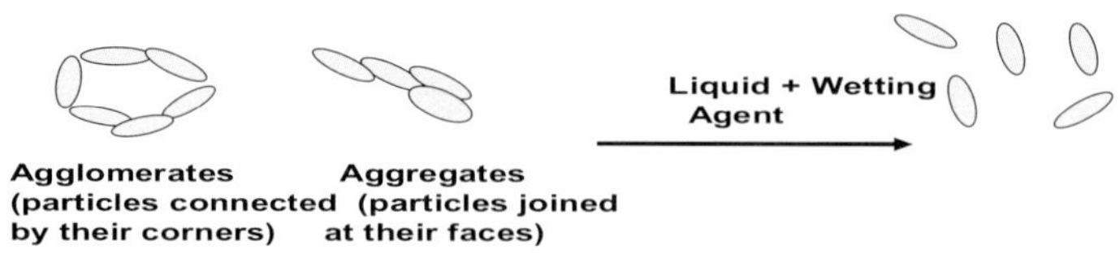

Figure 3.11 Schematic representation of aggregates and agglomerates.

the use of surface-active agents (wetting agents) of the ionic or nonionic type, which are capable of diffusing quickly (i.e. lower the dynamic surface tension) to the solid/liquid interface and displace the air entrapped by rapid penetration through the channels between the particles and inside any 'capillaries'. For wetting of hydrophobic powders into water, anionic surfactants, e.g. alkyl sulfates or sulfonates or nonionic surfactants of alcohol or alkylphenol ethoxylates are usually used.

A useful concept for choosing wetting agents of ethoxylated surfactants (see below) is the hydrophilic–lipophilic balance (HLB) concept,

$$HLB = \frac{\%\text{ of hydrophilic groups}}{5} \tag{28}$$

Most wetting agents of this class have an HLB number in the range 7–9.

The process of wetting of a solid by a liquid involves three types of wetting: adhesion wetting, W_a; immersion wetting, W_i; and spreading wetting, W_s. However, one can consider the work of dispersion wetting W_d as simply resulting from replacement of the solid/vapor interface with the solid/liquid interface.

Dispersion wetting W_d is given by the product of the external area of the powder A and the difference between γ_{SL} and γ_{SV}:

$$W_d = A(\gamma_{SL} - \gamma_{SV}) \tag{29}$$

Using Young's equation:

$$W_d = -A\gamma_{LV} \cos\theta \tag{30}$$

Thus wetting of the external surface of the powder depends on the liquid surface tension and contact angle. If $\theta < 90°$ then $\cos\theta$ is positive and the work of dispersion is negative, i.e. wetting is spontaneous. The most important parameter that determines wetting of the powder in the dynamic surface tension, $\gamma_{dynamic}$ (i.e. the value at short times). As will be discussed later, $\gamma_{dynamic}$ depends both on the diffusion coefficient of the surfactant molecule and on its concentration. Since wetting agents are added in sufficient amounts ($\gamma_{dynamic}$ is lowered sufficiently), spontaneous wetting is the rule rather than the exception.

Wetting of the internal surface requires penetration of the liquid into channels between and inside the agglomerates. The process is similar to forcing a liquid through fine capillaries. To force a liquid through a capillary with radius r, a pressure p is required that is given by

$$p = -\frac{2\gamma_{LV}\cos\theta}{r} = \frac{-2(\gamma_{SV} - \gamma_{SL})}{r\gamma_{LV}} \tag{31}$$

γ_{SL} has to be made as small as possible; rapid surfactant adsorption on the solid surface, i.e. low θ. When $\theta = 0$, $p \propto \gamma_{LV}$. Hence for penetration into pores, one requires a high γ_{LV}. Thus, wetting of the external surface requires a low contact angle θ and low surface tension γ_{LV}. Wetting of the internal surface (i.e. penetration through pores) requires low θ but high γ_{LV}. These two conditions are incompatible and a compromise has to be made: $\gamma_{SV} - \gamma_{SL}$ must be kept at a maximum. γ_{LV} should be kept as low as possible but not too low.

The above conclusions illustrate the problem of choosing the best wetting agent for a particular powder. This requires measurement of the above parameters and also testing the efficiency of the dispersion process.

3.1.13
Rate of Penetration of Liquids. The Rideal–Washburn Equation

For horizontal capillaries (gravity neglected), the depth of penetration l in time t is given by the Rideal–Washburn equation [15, 16]:

$$l^2 = \left(\frac{r\gamma_{LV}\cos\theta}{2\eta}\right)t \tag{32}$$

To enhance the rate of penetration, γ_{LV} has to be made as high as possible and θ and η as low as possible.

For dispersion of powders into liquids, one should use surfactants that lower θ while not reducing γ_{LV} too much. The viscosity of the liquid should also be kept at a minimum. Thickening agents (such as polymers) should not be added during the dispersion process. It is also necessary to avoid foam formation during the dispersion process.

For a packed bed of particles, r may be replaced by r/k^2, where r is the effective radius of the bed and k is the tortuosity factor, which takes into account the complex path formed by the channels between the particles, i.e.

$$l^2 = \left(\frac{r\gamma_{LV}\cos\theta}{2\eta k^2}\right)t \tag{33}$$

Thus a plot of l^2 versus t gives a straight line and from the slope of the line one can obtain θ.

The Rideal–Washburn equation can be applied to obtain the contact angle of liquids (and surfactant solutions) in powder beds; k should first be obtained using a liquid that produces a zero contact angle. This is discussed below.

3.1.14
Measurement of Wettability of Powders

3.1.14.1 Submersion Test – Sinking Time or Immersion Time

This by far the simplest (but qualitative) method for assessment of wettability of a powder by a surfactant solution. The time for which a powder floats on the surface of a liquid before sinking into the liquid is measured. A 100 ml volume of the surfactant solution is placed in a 250 ml beaker (of internal diameter 6.5 cm) and, after standing for 30 min, 0.30 g of loose powder (previously screened through a 200-mesh sieve) is distributed with a spoon over the surface of the solution. The time t for the 1–2 mm thin powder layer to disappear completely from the surface is measured using a stop-watch. Surfactant solutions with different concentrations are used and t is plotted versus surfactant concentration, as illustrated in Figure 3.12.

It can be seen from Figure 3.12 that the sinking time starts to decrease sharply above a critical surfactant concentration, reaching a minimum above this concentration. This procedure can be used to select the most effective wetting agent. The lower the surfactant concentration above which a rapid decrease in sinking time occurs and the lower the minimum wetting time obtained above this concentration, the more effective the wetter is.

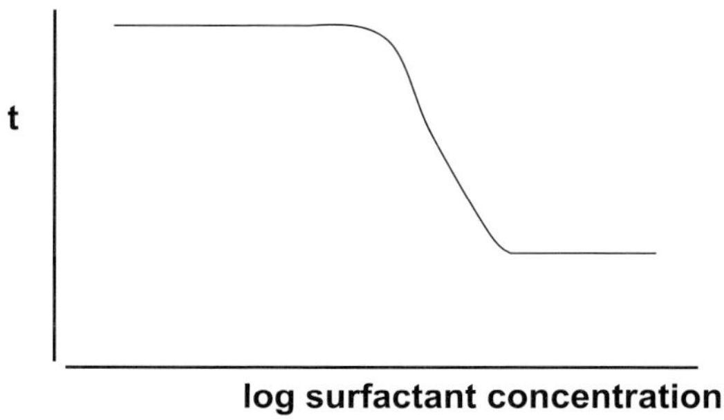

Figure 3.12 Sinking time as a function of surfactant concentration.

3.1.14.2 Measurement of Contact Angles of Liquids and Surfactant Solutions on Powders

As discussed above, the contact angle θ can be used for quantitative assessment of a surfactant as wetting agent for a particular powder. The simplest procedure is to measure the contact angle on a flat surface of the powder. This requires the preparation of a flat surface, for example by using a large crystal of the chemical or by compressing the powder to a thin plate (using high pressure, as commonly used, for example, for IR measurements). However, this procedure is inaccurate since on compressing the powder its surface will change and the measured contact angle will not be representative of the powder in question. However, this procedure may be used to compare various wetting agents and the assumption is made that the lower the surfactant concentration required to reach a zero contact angle, the more effective the wetter is.

The contact angle on powders can be measured more accurately by determining the rate of liquid penetration through a carefully packed bed of powder placed in a tube with sintered glass at the end (to retain the powder particles). It is essential to pack the powder uniformly in the tube (a plunger may be used in this case). By plotting l^2 (where l is the distance covered by the liquid flowing under capillary pressure) versus time t, a straight line is obtained (equation 33), and its slope is equal to $r\gamma_{LV} \cos \theta / 2\eta k^2$ (where r is the equivalent capillary radius, k is the tortuosity factor, γ_{LV} is the liquid surface tension and η is the liquid viscosity). This is illustrated in Figure 3.13; from the slope, $\cos \theta$ is obtained provided that r/k^2 is known.

The tortuosity factor k and the ratio r/k^2 can be obtained by using a liquid that completely wets the powder, giving a zero contact angle and $\cos \theta = 1$. The powder is carefully packed in a tube with sintered glass at the end using a specially designed cell fitted with a plunger for packing the powder (as supplied by Kruss). The cell is placed on top of liquid hexane, which gives a zero contact angle with most powders. The rate of penetration of hexane through the powder plug is measured by following the increase in weight ΔW of the cell with time. From the plot of ΔW^2 versus t one can obtain r/k^2 from the slope of the linear line. The cell is then removed and the hexane is allowed to evaporate completely.

The same cell with its powder packing is the placed on surfactant solutions with various concentrations and this allows one to obtain the contact angle as a function of concentration. The most effective wetter will be the one that gives $\theta = 0°$ at the lowest concentration.

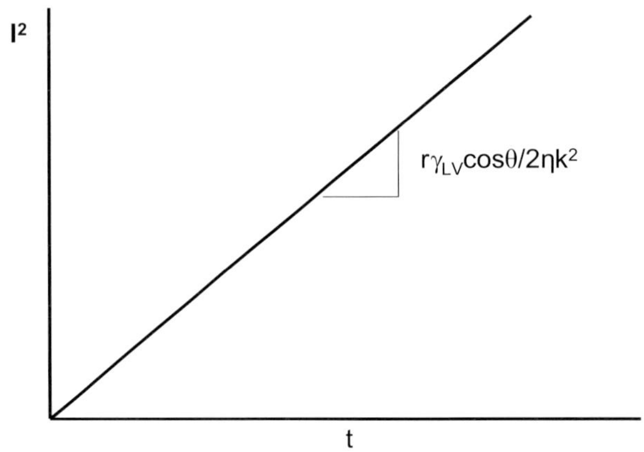

Figure 3.13 Variation of l^2 with t.

3.1.15
Wetting Agents for Hydrophobic Pigments

The most effective wetting agent is the one that gives a zero contact angle at the lowest concentration. For $\theta = 0°$ or $\cos \theta = 1$, γ_{SL} and γ_{LV} has to be as low as possible. This requires quick reduction of γ_{SL} and γ_{LV} under dynamic conditions during powder dispersion (this reduction should normally be achieved in less than 20 s). This requires fast adsorption of the surfactant molecules at both the liquid/vapor and solid/liquid interfaces.

It should be mentioned that a decrease in γ_{LV} is not always accompanied by a simultaneous decrease in γ_{SL}, hence it is necessary to have information on both interfacial tensions, which means that measurement of the contact angle is essential in the selection of wetting agents. Measurement of γ_{SL} and γ_{LV} should be carried out under dynamic conditions (i.e. at very short times). In the absence of such measurements, the sinking time described above could be applied as a guide for wetting agent selection.

The most commonly used wetting agents for hydrophobic solids are anionic or nonionic surfactants. To achieve rapid adsorption, the wetting agent should be either a branched chain with a central hydrophilic group or a short hydrophobic chain with a hydrophilic end group. The most commonly used wetting agent is Aerosol OT (diethylhexyl sulfosuccinate):

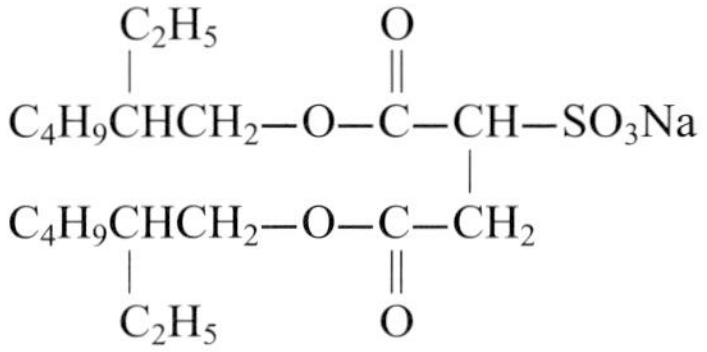

Structure 3.1

This molecule has a low critical micelle concentration (CMC) of 0.7 g dm^{-3}, and at and above the CMC the water surface tension is reduced to ~25 mN m^{-1} in less than 15 s.

An alternative anionic wetting agent is sodium dodecylbenzene sulfonate with a branched alkyl chain:

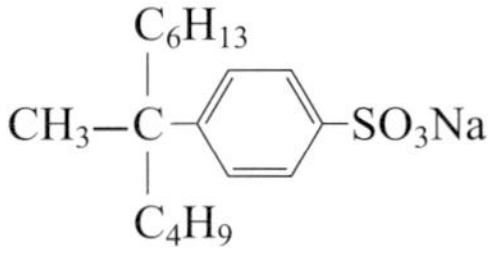

Structure 3.2

This molecule has a higher CMC (1 g dm^{-3}) than Aerosol OT. It is also not as effective in lowering the surface tension of water, reaching a value of 30 mN m^{-1} at and above the CMC. It is therefore not as effective as Aerosol OT for powder wetting.

Several nonionic surfactants, such as alcohol ethoxylates, can also be used as wetting agents. These molecules consist of a short hydrophobic chain (mostly C_{10}) that is also branched. A medium-chain poly(ethylene oxide) (PEO) mostly consisting of six EO units or less is used. The above molecules also reduce the dynamic surface tension within a short time (<20 s) and they have reasonably low CMCs. In all cases, one should use the minimum amount of wetting agent to avoid interference with the dispersant that needs to be added to maintain the colloid stability during dispersion and on storage.

3.1.16
Adsorption and Wetting Dynamics

Most processes of powder wetting work under dynamic conditions and improvement of their efficiency requires the use of surfactants that lower the liquid surface tension γ_{LV} under these dynamic conditions. The interfaces involved (particles separated from aggregates or agglomerates) are freshly formed and have only a small effective age of a few seconds or even less than 1 ms.

The most frequently used parameter to characterize the dynamic properties of liquid adsorption layers is the dynamic surface tension (which is a time-dependent quantity). Techniques should be available to measure γ_{LV} as a function of time (ranging from a fraction of a millisecond to minutes and hours or days).

To optimize the use of surfactants, specific knowledge of their dynamic adsorption behavior rather than equilibrium properties is of great interest [17]. It is necessary, therefore, to describe the dynamics of surfactant adsorption at a fundamental level.

3.1.17
General Theory of Adsorption Kinetics

The first physically sound model for adsorption kinetics was derived by Ward and Tordai [18]. It is based on the assumption that the time dependence of surface or interfacial tension, which is directly proportional to the surface excess Γ (mol m^{-2}), is caused by diffusion and transport of surfactant molecules to the interface. This is referred to as 'the diffusion-controlled adsorption kinetics model'. The interfacial surfactant concentration at any time t, $\Gamma(t)$, is given by the following expression:

$$\Gamma(t) = 2\left(\frac{D}{\pi}\right)^{\frac{1}{2}} c_0 t^{\frac{1}{2}} - \int_0^{t^{\frac{1}{2}}} c(0, t-\tau)\mathrm{d}(\tau)^{\frac{1}{2}} \tag{34}$$

where D is the diffusion coefficient, c_0 is the bulk concentration and τ is the thickness of the diffusion layer.

The above diffusion-controlled model assumes transport by diffusion of the surface-active molecules to be the rate-controlled step. The so-called 'kinetic-controlled model' is based on the transfer mechanism of molecules from solution to the adsorbed state and vice versa [17].

A schematic picture of the interfacial region is given in Figure 3.14, which shows three main states: (i) adsorption when the surface concentration Γ is lower than the equilibrium value Γ_0, (ii) equilibrium state when $\Gamma = \Gamma_0$ and (iii) desorption when $\Gamma > \Gamma_0$.

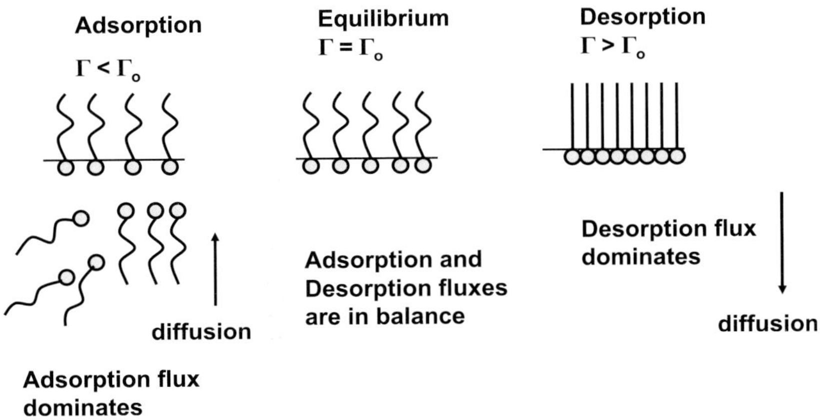

Figure 3.14 Representation of the fluxes of adsorbed surfactant molecules in the absence of liquid flow.

The transport of surfactant molecules from the liquid layer adjacent to the interface (subsurface) is simply determined by molecular movements (in the absence of forced liquid flow). At equilibrium, i.e. when $\Gamma = \Gamma_0$, the flux of adsorption is equal to the flux of desorption. Clearly, when $\Gamma < \Gamma_0$, the flux of adsorption predominates, whereas when $\Gamma > \Gamma_0$, the flux of desorption predominates [17].

In the presence of liquid flow, the situation becomes more complicated due to the creation of surface tension gradients [17]. These gradients, described by the Gibbs dilational elasticity [5], ε, initiate a flow of mass along the interface in the direction of the higher surface or interfacial tension (Marangoni effect), ε is given by

$$\varepsilon = A\frac{d\gamma}{dA} = \frac{d\gamma}{d\ln A} \tag{35}$$

where $d\gamma$ is the surface tension gradient and dA is the change in area of the interface. The above situation can arise, for example, if an adsorption layer is compressed or stretched, as illustrated in Figure 3.15.

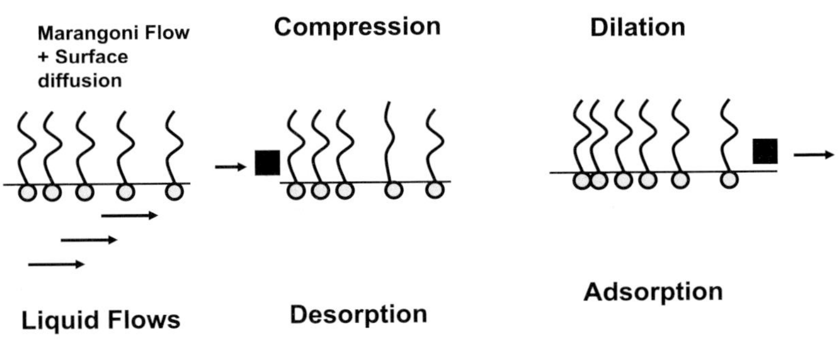

Figure 3.15 Representation of surfactant transport at the surface and in the bulk of a liquid.

A qualitative model that can describe adsorption kinetics is described by the following equation:

$$\Gamma(t) - c_0 \left(\frac{Dt}{\pi}\right)^{\frac{1}{2}} \tag{36}$$

Equation 36 gives a rough estimate and results from equation 34 when the second term on the right hand-side is neglected. An equation equivalent to equation 36 was derived by Panaitov and Petrov [19]:

$$c(0,t) = c_0 - \frac{2}{(D\pi)^{\frac{1}{2}}} \int_0^{t^{\frac{1}{2}}} \frac{\mathrm{d}\Gamma(t-\tau)}{\mathrm{d}t} \mathrm{d}\tau^{\frac{1}{2}} \tag{37}$$

Hansen [20] and Miller and Lunkenheimer [21] gave numerical solutions to the integrals of equations 34 and 36 and obtained a simple expression using a Langmuir isotherm:

$$\Gamma(t) = \Gamma_\infty \frac{c(0,t)}{a_\mathrm{L} + c(0,t)} \tag{38}$$

where a_L is the constant in the Langmuir isotherm (mol m^{-3}). The corresponding equation for the variation of surface tension γ with time is as follows (Langmuir–Szyszowski equation):

$$\gamma = \gamma_0 + RT\Gamma_\infty \ln\left[1 - \frac{\Gamma(t)}{\Gamma_\infty}\right] \tag{39}$$

Calculations based on equations 37–39 are given in Figure 3.16, with different values of $c_0/\mathrm{a_L}$ [20].

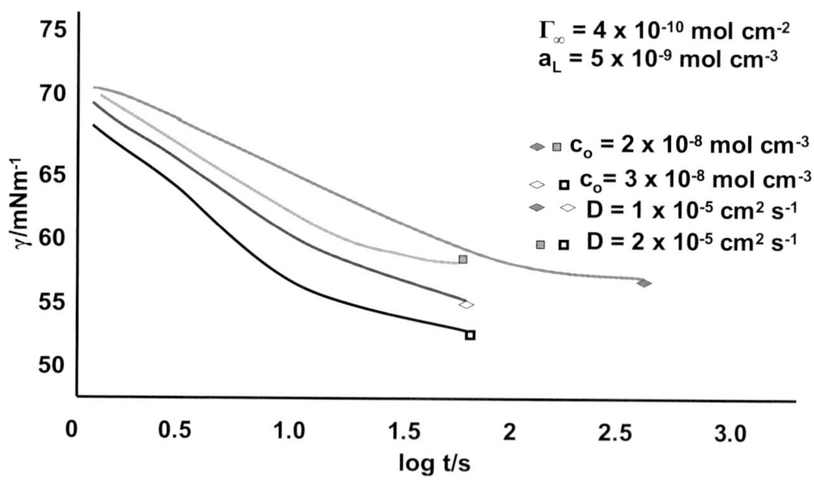

Figure 3.16 Surface tension γ–log t curves calculated on the basis of equations 37–39.

3.1.17.1 Adsorption Kinetics from Micellar Solutions

Surfactants form micelles above the critical micelle concentration (cmc) of different sizes and shapes, depending on the nature of the molecule, temperature, electrolyte concentration, etc. The dynamic nature of micellization can be described by two main relaxation processes, τ_1 (the lifetime of a monomer in a micelle) and τ_2 (the lifetime of the micelle, i.e. complete dissolution into monomers).

The presence of micelles in equilibrium with monomers influences the adsorption kinetics remarkably. After a fresh surface has been formed, surfactant monomers are adsorbed, resulting in a concentration gradient of these monomers. This gradient will be equalized by diffusion to re-establish a homogeneous distribution. Simultaneously, the micelles are no longer in equilibrium with monomers within the range of the concentration gradient. This leads to a net process of micelle dissolution or rearrangement to re-establish the local equilibrium. As a consequence, a concentration gradient of micelles results, which is equalized by diffusion of micelles [17].

Based on the above concepts, one would expect that the ratio of monomers c_1 to micelles c_m, the aggregation number n, rate of micelle formation k_f and rate of micelle dissolution k_d will influence the rate of the adsorption process. A schematic picture of the kinetic process in the presence of micelles is given in Figure 3.17. This picture shows that to describe the kinetics of adsorption, one must take into account the diffusion of monomers and micelles in addition to the kinetics of micelle formation and dissolution.

Several processes may take place and these are represented schematically in Figure 3.18. Three main mechanisms may be considered, namely formation–dissolution (Figure 3.18a), rearrangement (Figure 3.18b) and stepwise aggregation–dissolution (Figure 3.18c). To describe the effect of micelles on the adsorption kinetics, one should know several parameters, such as micelle aggregation number and rate constants of micelle kinetics [22].

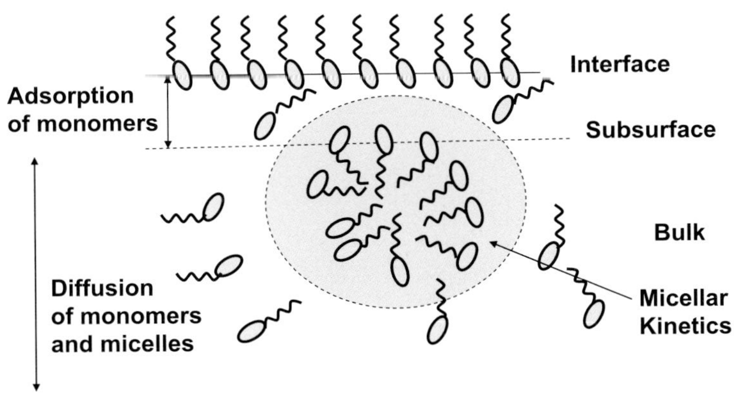

Figure 3.17 Representation of the adsorption process from a micellar solution.

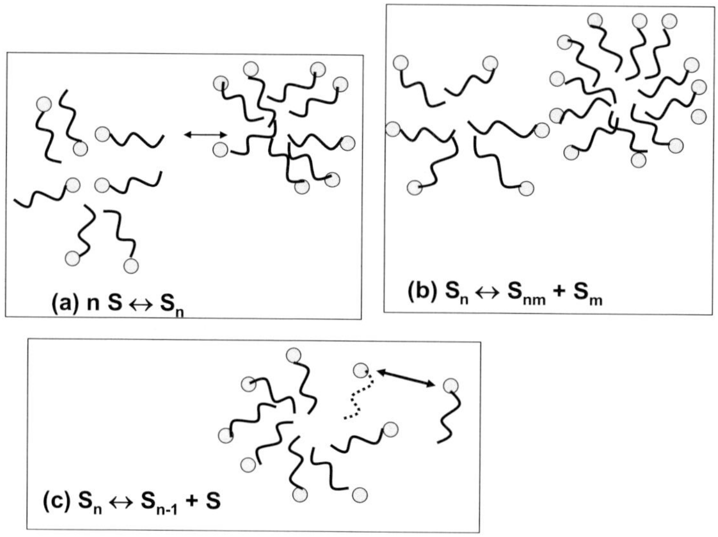

Figure 3.18 Scheme of micelle kinetics.

3.1.17.2 Experimental Techniques for Studying Adsorption Kinetics

The two most suitable techniques for studying adsorption kinetics are the drop volume method and the maximum bubble pressure method. The former method can obtain information on adsorption kinetics in the range of seconds to some minutes. However, it has the advantage of measurement both at the air/liquid and liquid/liquid interfaces. The maximum bubble pressure method allows one to obtain measurements in the millisecond range, but it is restricted to the air/liquid interface. Below a description of these two techniques is given.

The Drop Volume Technique A schematic representation of the drop volume apparatus [23] is given in Figure 3.19.

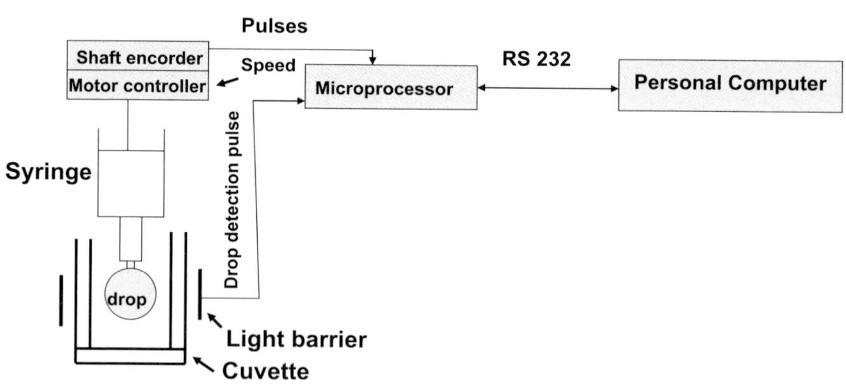

Figure 3.19 Representation of the drop volume apparatus.

A metering system in the form of a motor-driven syringe allows the formation of the liquid drop at the tip of a capillary, which is positioned in a sealed cuvette. The cuvette is either filled with a small amount of the measuring liquid, to saturate the atmosphere, or with a second liquid in the case of interfacial studies. A light barrier arranged below the forming drop permits the detection of drop detachment from the capillary. Both the syringe and the light barriers are computer controlled and allow fully automatic operation of the set-up. The syringe and the cuvette are temperature controlled by a water jacket, which makes interfacial tension measurements possible in the temperature range 10–90 °C.

As mentioned above, the drop volume method is of dynamic character and it can be used for adsorption processes in the time interval from seconds to some minutes. At a short drop time, the so-called hydrodynamic effect has to be considered [24]. This gives rise to apparently higher surface tension. Kloubek *et al.* [25] used an empirical equation to account for this effect:

$$V_e = V(t) - \frac{K_v}{t} \tag{40}$$

where V_e is the unaffected drop volume, $V(t)$ is the measured drop volume and K_v is a proportionality factor that depends on surface tension γ, density difference $\Delta\rho$ and tip radius r_{cap}.

Miller and co-workers [17] obtained the following equation for the variation of drop volume $V(t)$ with time:

$$V(t) = V_e + t_0 V_e + t_0 F = V_e \left(1 + \frac{t_0}{t - t_0}\right) \tag{41}$$

where F is the liquid flow per unit time and is given by

$$F = \frac{V(t)}{t} = \frac{V_e}{t - t_0} \tag{42}$$

The drop volume technique is limited in its application. Under conditions of fast drop formation and larger tip radii, the drop formation shows irregular behavior.

Maximum Bubble Pressure Technique This is the most useful technique for measuring adsorption kinetics at short times, particularly if correction for the so-called 'dead time', τ_d, is made. The dead time is simply the time required to detach the bubble after it has reached its hemispherical shape. A schematic representation of the principle of maximum bubble pressure is shown in Figure 3.20, which shows the evolution of a bubble at the tip of a capillary. The figure also shows the variation of pressure p in the bubble with time.

At $t = 0$ (initial state), the pressure is low (note that the pressure is equal to $2\gamma/r$; since r of the bubble is large p is small). At $t = \tau$ (smallest bubble radius that is equal to the tube radius), p reaches a maximum. At $t = \tau_b$ (detachment

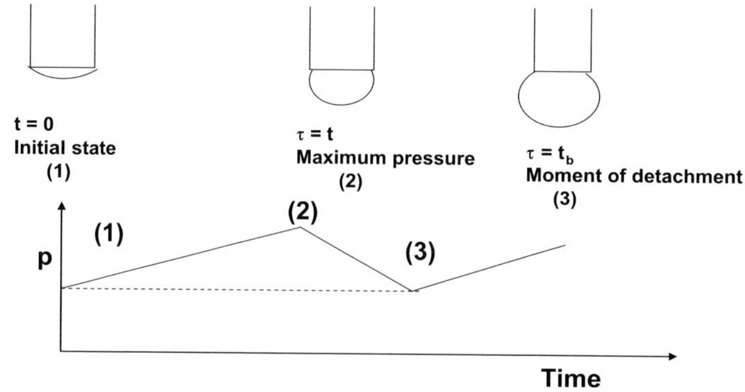

Figure 3.20 Scheme of bubble evolution and pressure change with time.

time), *p* decreases since the bubble radius increases. The design of a maximum bubble pressure method for high bubble formation frequencies (short surface age) requires the following: (i) measurement of bubble pressure, (ii) measurement of bubble formation frequency and (iii) estimation of surface lifetime and effective surface age. The first problem can easily be solved if the system volume (which is connected to the bubble) is large enough in comparison with the bubble separating from the capillary. In this case, the system pressure is equal to the maximum bubble pressure. The use of an electric pressure transducer for measuring bubble formation frequency presumes that pressure oscillations in the measuring system are distinct enough and this satisfies point (ii). Estimation of the surface lifetime and effective surface age, i.e. point (iii), requires estimation of the dead time τ_d.

A schematic representation of the set-up for measuring the maximum bubble pressure and surface age is shown in Figure 3.21. The air coming from a microcompressor first flows through the flow capillary. The air flow rate is determined

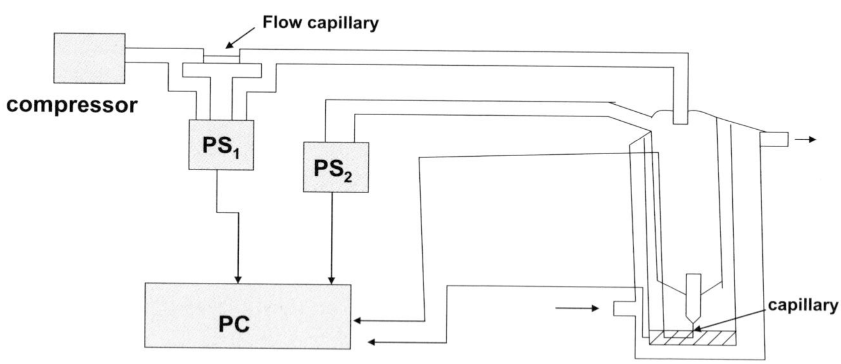

Figure 3.21 Maximum bubble pressure apparatus.

by measuring the pressure difference at both ends of the flow capillary with the electric transducer PS1. Thereafter, the air enters the measuring cell and the excess air pressure in the system is measured by a second electric sensor PS2. In the tube which leads the air to the measuring cell, a sensitive microphone is placed. The measuring cell is equipped with a water jacket for temperature control, which simultaneously holds the measuring capillary and two platinum electrodes, one of which is immersed in the liquid under study and the other is situated exactly opposite to the capillary and controls the size of the bubble. The electric signals from the gas flow sensor PS1 and pressure transducer PS2, the microphone and the electrodes, and also the compressor are connected to a personal computer which operates the apparatus and acquires the data.

The value of τ_d, equivalent to the time interval necessary to form a bubble of radius R, can be calculated using Poiseuille's law:

$$\tau_d = \frac{\tau_b L}{Kp}\left(1 + \frac{3r_{ca}}{2R}\right) \tag{43}$$

where K is given by

$$K = \frac{\pi r^4}{8\eta l} \tag{44}$$

where η is the gas viscosity, l is the length, L is the gas flow rate and r_{ca} is the radius of the capillary.

The calculation of the dead time τ_d can be simplified by taking into account the existence of two gas flow regimes for the gas flow leaving the capillary: the bubble flow regime when $\tau > 0$ and the jet regime when $\tau = 0$ and hence $\tau_b = \tau_d$. A typical dependence of p on L is shown in Figure 3.22.

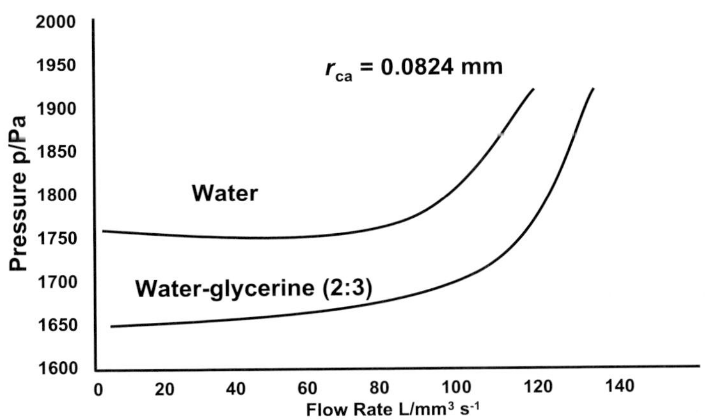

Figure 3.22 Dependence of p on gas flow rate L at 30 °C.

On the right of the critical point, the dependence of p on L is linear in accordance with Poiseuille's law. Under these conditions,

$$\tau_d = \tau_b \frac{L p_c}{L_c p} \tag{45}$$

where L_c and p_c are related to the critical point and L and p are the actual values of the dependence on the left of the critical point.

The surface lifetime can be calculated from

$$\tau = \tau_b - \tau_d = \tau_b \left(1 - \frac{L p_c}{L_c p}\right) \tag{46}$$

The critical point in the dependence of p and L can be easily located and is included in the software of the computer program.

The surface tension value in the maximum bubble pressure method is calculated using the Laplace equation:

$$p = \frac{2\gamma}{r} + \rho h g + \Delta p \tag{47}$$

where ρ is the density of the liquid, g is the acceleration due to gravity, h is the depth to which the capillary is immersed in the liquid and Δp is a correction factor to allow for hydrodynamic effects.

3.2 Breaking of Aggregates and Agglomerates (Deagglomeration)

As mentioned above, all pigments are supplied as powders consisting of aggregates (where the particles are connected by their surfaces) or agglomerates (where the particles are connected by their corners). For example, pigmentary titanium dioxide mostly exists in powder form as loose agglomerates several tens of micrometers in diameter. These pigments are surface coated by the manufacturer for two main reasons. First, the surface coating reduces the cohesive forces of the powder, thus assisting the deagglomeration process. Second, the coating (SiO_2 and Al_2O_3) deactivates the surface rutile pigment (by reducing the photochemical activity), which otherwise accelerates the degradation of the resin on weathering.

The 'grinding stage' in mill-based manufacture is not a comminution stage but a dispersion process of the pigment agglomerates, whereby the latter are separated into 'single' primary particles. However, some of the primary particles may consist of sinters of TiO_2 crystals produced during the surface coating stage. To separate the particles in an aggregate or agglomerate, one requires the use of a wetting/dispersing system. As mentioned above, the wetting agent, which is usually a short-chain surfactant molecule, can seldom prevent the reaggregation of the primary particles after the dispersion process. Therefore, to prevent the reaggregation of particles, a dispersing agent is required. The dispersing agent may replace the wetting agent at the solid/liquid interface or become co-adsorbed with the wetting agent. The dispersant produces an effective repulsive barrier on close approach of the particles (see Chapter 4). This repulsive barrier is particularly important for concentrated pigment dispersions (which may contain more than 50% by volume of solids).

The main criteria for an effective dispersant (which will be discussed in detail in Chapter 4) are as follows. (i) *Strong adsorption* or 'anchoring' to the particle surface. (ii) *A high repulsive barrier*: the stabilizing chain A of the dispersant must provide an effective repulsive barrier to prevent flocculation by van der Waals attraction; three main mechanisms of stabilization can be considered, electrostatic, as produced, for example, by ionic surfactants, steric, as produced by nonionic polymeric surfactants of +A–B, B–A–B, A–B–A or AB_n graft copolymers (where A is the 'anchor' chain and B is the 'stabilizing' chain), and electrosteric, as produced by polyelectrolytes. (iii) *Strong solvation of the stabilizing B chain*, that is, it should be in a good solvent condition, i.e. very soluble in the medium and strongly solvated by its molecules. Solvation of the chain by the medium is determined by the chain–solvent (Flory–Huggins) interaction parameter χ. In good solvent conditions, $\chi < 0.5$ and hence the mixing or osmotic interaction is positive (repulsive). χ should be maintained at <0.5 under all conditions, e.g. low and high temperature, in the presence of electrolytes and other components of the formulation such as addition of anti-freeze (mostly propylene glycol). (iv) *Reasonably thick adsorbed layer*: the adsorbed layer thickness of the B chains, usually described by a hydrodynamic value δ_h (i.e. the thickness δ plus any contribution from the solvation shell), should be sufficiently large to prevent the formation of a deep minimum which may result in flocculation (although reversible) and an increase in the viscosity of the suspension. A value of $\delta_h > 5$ nm is usually sufficient to prevent the formation of a deep minimum.

3.2.1 Classification of Dispersants

(i) Surfactants Different types are as follows: anionic, e.g. sodium dodecyl sulfate (SDS), $C_{12}H_{25}OSO_3Na$, and sodium dodecylbenzene sulfonate (NaDBS), $C_{12}H_{25}C_6H_4SO_3Na$; cationic, e.g. dodecyltrimethylammonium chloride, $C_{12}H_{25}N(CH_3)_3Cl$; amphoteric, e.g. betaines such as laurylamidopropyldimethylbetaine, $C_{12}H_{25}CON(CH_3)_2CH_2COOH$; and nonionic, the most common being the alcohol ethoxylates $RO(CH_2CH_2O)_nH$, e.g. $C_{13/15}(EO)_n$ with n being 7, 9, 11 or 20. These surfactants are not the most effective dispersants since the adsorption by the $C_{13/15}$ chain is not very strong. To enhance the adsorption on hydrophobic surfaces, a polypropylene oxide (PPO) chain is introduced in the molecule, giving $RO(PPO)_m(PEO)_nH$. A more effective nonionic surfactant with a strong adsorption is obtained by using a tristyrylphenol with PEO, e.g.

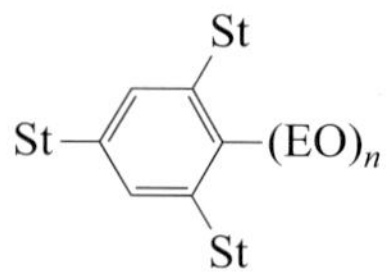

Structure 3.3

The tristyrylphenol hydrophobic chain adsorbs strongly on a hydrophobic surface. This is due to the poor 'anchoring' of the chain to the surface and the high solubility of the polymer in water.

(ii) Polymeric Surfactants Homopolymers consisting of the same repeating units such as poly(ethylene oxide) (PEO) or polyvinylpyrrolidone are not good dispersants for hydrophobic solids in aqueous media, due to the poor 'anchoring' of the chain to the surface and the high solubility of the polymer in water. In most cases these homopolymers do not adsorb at all on the particles, since the decrease in configurational entropy on adsorption is not compensated for by an adsorption energy (the adsorption energy per segment χ_s is very low).

As mentioned above, the most effective polymeric surfactants as dispersants are those of the A–B, B–A–B block and AB_n or BA_n graft types. The 'anchor chain' A is chosen to be highly insoluble in the medium and has a strong affinity to the surface. Examples of A chains for hydrophobic solids are polystyrene (PS), poly(methyl methacrylate) (PMMA), poly(propylene oxide) (PPO) and alkyl chains provided that they have several attachments to the surface. The stabilizing chain B has to be soluble in the medium and strongly solvated by its molecules. The B chain–solvent interaction should be strong, giving a Flory–Huggins χ parameter < 0.5 under all conditions. Examples of B chains are poly(ethylene oxide) (PEO), poly(vinyl alcohol) (PVA) and polysaccharides (e.g. polyfructose). Several examples of commercially available B–A–B block copolymers are available, such as B–A–B block copolymers of PEO and PPO: Pluronics. Several molecules of PEO–PPO–PEO are available with various proportions of PEO and PPO. The commercial name is followed by a letter L (Liquid), P (Paste) or F (Flake). This followed by two numbers that represent the composition; the first digit represents the PPO molecular mass and the second represents the percentage of PEO, e.g. Pluronic F68 (PPO molecular mass 1508–1800 + 80% or 140 mol EO) and Pluronic L62 (PPO molecular mass 1508–1800 + 20% or 15 mol EO). In many cases, two Pluronics with high and low EO contents are used together to enhance the dispersing power.

Graft copolymers of the AB_n type are also available, e.g. an AB_n graft copolymer based on a poly(methyl methacrylate) (PMMA) backbone [with some poly(methacrylic acid)] on which several PEO chains (with an average molecular mass of 750) are grafted:

```
~~~~~~~PMMA~~~~~~~~
  |       |    |    |
PEO     PEO  PEO  PEO
```

Structure 3.4

This is a very effective dispersant particularly for suspensions with a high solids content. The graft copolymer is strongly adsorbed on hydrophobic surfaces with several attachment points by the small PMMA loops of the backbone and a strong steric barrier is obtained by the highly hydrated PEO chains in aqueous solutions.

A novel BA_n graft has recently been synthesized, namely INUTEC SP1 (Orafti, Belgium) consisting of inulin, a linear polyfructose chain A (with degree of polymerization > 23), on which several alkyl chains have been grafted (Figure 3.23). The polymeric surfactant adsorbs with multi-point attachment with several alkyl chains.

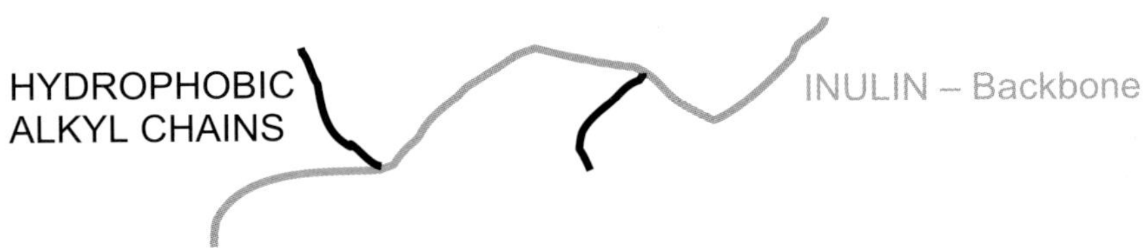

Figure 3.23 INUTEC SP1.

Another commercially available 'blocky' copolymer is partially hydrolyzed poly(vinyl acetate) (PVAc), referred to commercially as poly(vinyl alcohol) (PVA). The molecule contains short blocks of PVAc which form the anchor chains to the hydrophobic surface, leaving several loops and tails of PVA chains which are strongly hydrated to give an effective steric barrier:

$$-(CH_2{-}\underset{\displaystyle OH}{\underset{|}{C}H})_x{-}(CH_2{-}\underset{\displaystyle OCOCH_3}{\underset{|}{C}H})_y{-}(CH_2{-}\underset{\displaystyle OH}{\underset{|}{C}H})_x{-}(CH_2{-}\underset{\displaystyle OH}{\underset{|}{C}H})_x{-}(CH_2{-}\underset{\displaystyle OCOCH_3}{\underset{|}{C}H})_y{-}$$

Structure 3.5

Several commercial PVA grades are available with molecular weights in the range 20 000–100 000 and acetate content in the range 4–12%. The molecule is designated by two numbers representing the degree of hydrolysis and viscosity of a 4% solution (which gives a rough estimate of molecular weight). For example, Moviol 88/10 refers to a degree of hydrolysis of 88% (12% acetate groups) and a viscosity of 10 mPa s of a 4% solution.

(iii) Polyelectrolytes In naphthalene formaldehyde sulfonated condensate:

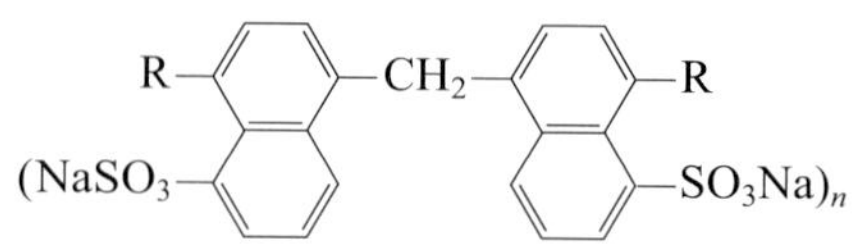

Structure 3.6

n varies between 2 and 9 units, i.e. the molecule has a wide distribution of molecular weights.

Another commercially available dispersant type is the lignosulfonates, which are isolated from the waste liquor from wood pulping by the sulfite process, during which lignin is sulfonated. They are also produced by sulfonating lignin by alkaline pulping of wood by the Kraft process. Lignosulfonates as dispersants are mixtures of polyelectrolytes with a molecular weigh ranging from 2000 to 10 000. The exact structure of lignosulfonates is not completely known, but guaiacylpropyl groups with the sulfate groups attached to the aliphatic chains of lignin have been identified. The degree of sulfonation varies from 0.3 to 1.0 per phenyl unit. The commercial products, namely Polyfon (Wesvaco) and Ufoxane (Borregard), are described by the degree of sulfonation per 840 units of lignin. For example, Polyfon H has a degree of sulfonation of 0.5, whereas Polyfon T has a degree of sulfonation of 2.0. The most effective lignosulfonates for hydrophobic solids in aqueous solution are those with lower degrees of sulfonation, which give higher adsorption.

3.2.2 Assessment and Selection of Dispersants

3.2.2.1 Adsorption Isotherms

These are by far the most quantitative methods for the assessment of the dispersing power. Known amounts of solids (m g) with a surface area A ($m^2\ g^{-1}$) are equilibrated at constant temperature with dispersant solutions with various concentrations C_1. The bottles containing the various dispersions are rotated for several hours until equilibrium is reached. The particles are removed from the dispersant solution by centrifugation and/or filtration through Millipore filters. The dispersant concentration in the supernatant liquid C_2 is determined analytically by a suitable technique that can measure low concentrations.

The amount of adsorption Γ ($mg\ m^{-2}$ or $mol\ m^{-2}$) is calculated:

$$\Gamma = \frac{C_1 - C_2}{mA} \qquad (48)$$

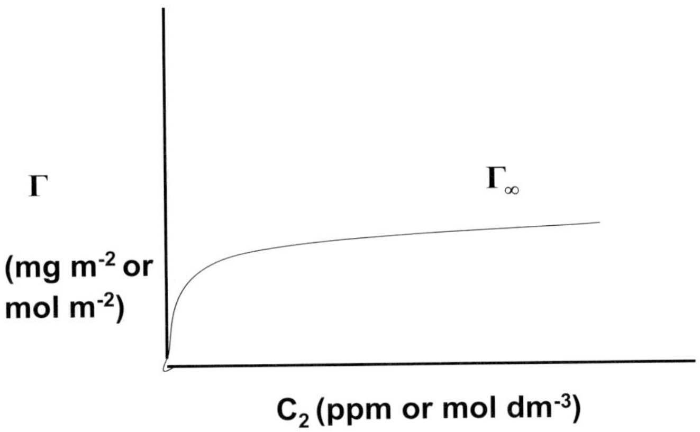

Figure 3.24 Langmuir-type adsorption isotherm.

A plot of Γ versus C_2 gives the adsorption isotherm. Two types of isotherms can be distinguished: a Langmuir type for reversible adsorption of surfactants (Figure 3.24) and a high-affinity isotherm (Figure 3.25) for irreversible adsorption of polymeric surfactants.

In both cases a plateau adsorption value Γ_∞ is reached at a given value of C_2. In general, the value of Γ_∞ is reached at lower C_2 for polymeric surfactant adsorption compared with small molecules. The high-affinity isotherm obtained with polymeric surfactants implies that the first added molecules are virtually completely adsorbed and such a process is irreversible. The irreversibility of adsorption is checked by carrying out a desorption experiment. The suspension at the plateau value is centrifuged and the supernatant liquid is replaced with water. After redispersion, the suspension is centrifuged again and the concentration of the polymeric surfactant in the supernatant liquid is determined analytically. Owing to a lack of desorption, the above concentration will be very low, indicating that the polymer remains on the particle surface.

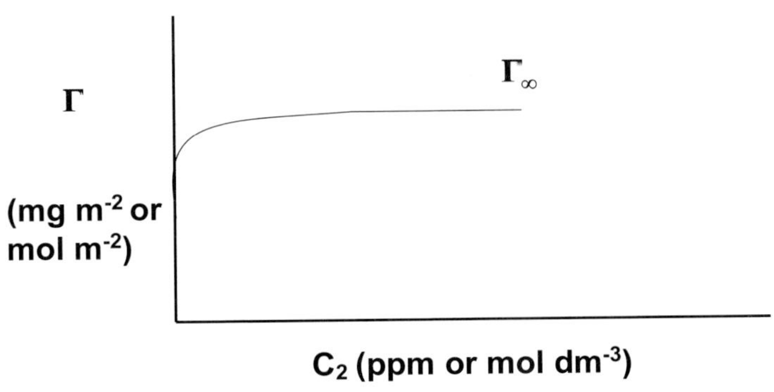

Figure 3.25 High-affinity isotherm.

3.2.3 Measurement of Dispersion and Particle Size Distribution

An effective dispersant should result in complete dispersion of the powder into single particles. In addition, on wet milling (comminution) a smaller particle distribution should be obtained. The efficiency of dispersion and reduction of particle size can be understood from the behavior of the dispersant. Strong adsorption and an effective repulsive barrier prevent any aggregation during the dispersion process. It is necessary in this case to include the wetter (which should be kept at the optimum concentration). Adsorption of the dispersant at the solid/liquid interface results in lowering of γ_{SL} and this reduces the energy required for breaking the particles into smaller units. In addition, by adsorption in crystal defects, crack propagation occurs (the Rehbinder effect), and this results in the production of smaller particles.

Several methods may be applied for measurement of the particle size distribution and these are summarized below.

3.2.3.1 Optical Microscopy

This is by far the most valuable tool for a qualitative or quantitative examination of the dispersion. Information on the size, shape, morphology and aggregation of particles can be conveniently obtained with a minimum time required for sample preparation. However, optical microscopy has some limitations: (i) the minimum size that can be detected – the practical lower limit for accurate measurement of particle size is 1.0 μm, although some detection may be obtained down to 0.3 μm; (ii) the image contrast may not be good enough for observation, particularly when using a video camera, which is mostly used for convenience. The contrast can be improved by decreasing the aperture of the iris diaphragm, but this reduces the resolution.

Three main attachments to the optical microscope are possible, as follows.

(i) Phase Contrast This utilizes the difference between the diffracted waves from the main image and the direct light from the light source. The specimen is illuminated with a light cone and this illumination is within the objective aperture. The light illuminates the specimen and generates zero-order and higher orders of diffracted light. The zero-order light beam passes through the objective and a phase plate, which is located at the objective back focal plane. The difference between the optical path of the direct light beam and that of the beam diffracted by a particle causes a phase difference. The constructive and destructive interferences result in brightness changes which enhance the contrast. This produces sharp images allowing one to obtain particle size measurements more accurately.

(ii) Differential Interference Contrast (DIC) This gives a better contrast than the phase contrast method. It utilizes a phase difference to improve contrast, but the separation and recombination of a light beam into two beams are accomplished by prisms. DIC generates interference colors and the contrast effects indicate the refractive index difference between the particle and medium.

(iii) Polarized Light Microscopy This illuminates the sample with linearly or circularly polarized light, in either a reflection or transmission mode. One polarizing element, located below the stage of the microscope, converts the illumination to polarized light. The second polarizer is located between the objective and the ocular and is used to detect polarized light.

The optical microscope can be used to observe dispersed particles and flocs. Particle sizing can be carried out using manual, semi-automatic or automatic image analysis techniques. In the manual method (which is tedious), the microscope is fitted with a minimum of 10× and 43× achromatic or apochromatic objectives equipped with high numerical apertures (10×, 15× and 20×), a mechanical *XY* stage, a stage micrometer and a light source. One of the difficulties with the evaluation of dispersions by optical microscopy is the quantification of data. The number of particles in at least six different size ranges must be counted to obtain a distribution. This problem can be alleviated by the use of automatic image analysis, which can also give an indication on the floc size and its morphology.

3.2.3.2 **Electron Microscopy**

This utilizes an electron beam to illuminate the sample. The electrons behave as charged particles, which can be focused by annular electrostatic or electromagnetic fields surrounding the electron beam. Due to the very short wavelength of electrons, the resolving power of an electron microscope exceeds that of an optical microscope by ~200-fold. The resolution depends on the accelerating voltage, which determines the wavelength of the electron beam and magnifications as high as 200 000 can be reached with intense beams, but this could damage the sample. Mostly the accelerating voltage is kept below 100–200 kV and the maximum magnification obtained is below 100 000×. Two main types of electron microscopes are used: transmission (TEM) and scanning (SEM).

TEM displays an image of the specimen on a fluorescent screen and the image can be recorded on a photographic plate or film. The sample is deposited on a Formvar (polyvinylformal) film resting on a grid to prevent charging of the simple. The sample is usually observed as a replica by coating with an electron-transparent material (such as gold or graphite). The preparation of the sample for the TEM may alter the state of dispersion and cause aggregation. Freeze fracturing techniques have been developed to avoid some of the alterations to the sample during sample preparation. Freeze fracturing allows the dispersions to be examined without dilution and replicas can be made of dispersions containing water. It is necessary to have a high cooling rate to avoid the formation of ice crystals.

SEM can show particle topography by scanning a very narrowly focused beam across the particle surface. The electron beam is directed normally or obliquely at the surface. The backscattered or secondary electrons are detected in a raster pattern and displayed on a monitor screen. The images provided by secondary electrons exhibit good three-dimensional detail. The backscattered electrons, reflected from the incoming electron beam, indicate regions of high electron density. Most

SEMs are equipped with both types of detectors. The resolution of the SEM depends on the energy of the electron beam, which does not exceed 30 kV, and hence the resolution is lower than that obtained with a TEM.

3.2.3.3 Confocal Scanning Laser Microscopy (CLSM)

CLSM is a very useful technique for the identification of dispersions. It uses a variable pinhole aperture or variable width slit to illuminate only the focal plane by the apex of a cone of laser light. Out-of-focus items are dark and do not distract from the contrast of the image. As a result of extreme depth discrimination (optical sectioning), the resolution is considerably improved (up to 40% compared with optical microscopy). The CLSM technique acquires images by laser scanning or uses computer software to subtract out-of-focus details from the in-focus image. Images are stored as the sample is advanced through the focal plane, which allows one to construct three-dimensional images.

3.2.3.4 Scattering Techniques

These are by far the most useful methods for the characterization of dispersions and in principle they can give quantitative information on the particle or droplet size distribution, floc size and shape. The only limitation of the methods is the need to use sufficiently dilute samples to avoid interferences, such as multiple scattering, which make the interpretation of the results difficult. However, backscattering methods have recently been designed to allow one to measure the sample without dilution. In principle one can use any electromagnetic radiation such as light, X-rays or neutrons, but in most industrial laboratories only light scattering is applied (using lasers).

Scattering techniques can be conveniently divided into the following classes: (i) time-average light scattering, static or elastic scattering; (ii) turbidity measurements, which can be carried out using a simple spectrophotometer; (iii) light diffraction technique; (iv) dynamic (quasi-elastic) light scattering, which is usually referred as photon correlation spectroscopy (PCS), which is a rapid technique that is very suitable for measuring submicron particles or droplets (nano-size range); and (v) backscattering technique, which is suitable for measuring concentrated samples.

Application of any of the above methods depends on the information required and the availability of the instrument.

Time-average Light Scattering In this method, the dispersion, which is sufficiently diluted to avoid multiple scattering, is illuminated by a collimated light (usually laser) beam and the time-average intensity of scattered light is measured as a function of scattering angle. Three regimes can be identified (a) The Rayleigh regime, where the particle radius R is smaller than $\lambda/20$ (λ being the wavelength of incident light): the scattering intensity is given by the equation

$$I(Q) = \text{instrument constant} \times \text{material constant} \times N{V_p}^2 \qquad (49)$$

where Q is the scattering vector, which depends on the wavelength of light used. The material constant depends on the difference between the refractive index of the particle or droplet and that of the medium. N is the number of particles or droplets and V_p is the volume of each particle or droplet. Assuming that the particles are spherical, one can obtain the average size using equation 49. (b) The Rayleigh–Gans–Debye (RGD) regime, where $\lambda/20 < R < \lambda$: this regime is more complicated than the Rayleigh regime and the scattering pattern is no longer symmetrical about the line corresponding to the 90° angle but favors forward scattering ($\theta < 90°$) or backscattering ($180° > \theta > 90°$). (c) The Mie regime, where $R > \lambda$: the scattering behavior is more complex than in the RGD regime and the intensity exhibits maxima and minima at various scattering angles depending on particle size and refractive index. The Mie theory of light scattering can be used to obtain the particle size distribution using numerical solutions. One can also obtain information on particle shape.

Turbidity Measurements (Total Light Scattering Technique) This can be used to measure particle size, flocculation and particle sedimentation. The technique is simple and easy to use (a single- or double-beam spectrophotometer or a nephelometer can be used). For non-absorbing particles, the turbidity τ is given by

$$\tau = \frac{1}{L} \ln\left(\frac{I_0}{I}\right) \tag{50}$$

where L is the pathlength, I_0 is the intensity of the incident beam and I is the intensity of the transmitted beam. The particle size measurement assumes that the light scattering by a particle is singular and independent of other particles Any multiple scattering complicates the analysis. According to the Mie theory, the turbidity is related to the particle number N and their cross section πr^2 (where r is the particle radius) by

$$\tau = Q\pi r^2 N \tag{51}$$

where Q is the total Mie scattering coefficient, which depends on the particle size parameter α (which depends on particle diameter and wavelength of incident light λ) and the ratio of the refractive indices of the particles and the medium m. Q depends on α in an oscillatory mode, exhibiting a series of maxima and minima whose positions depend on m. For particles with $R < (1/20)\lambda$, $a < 1$ and it can be calculated using the Rayleigh theory. For $R > \lambda$, Q approaches 2, and between these two extremes the Mie theory is used. If the particles are not monodisperse (as is the case with most practical systems), the particle size distribution must be taken into account. Using this analysis, one can establish the particle size distribution using numerical solutions.

Light Diffraction Techniques This is a rapid and non-intrusive technique for the determination of particle size distribution in the range 2–300 μm with good accuracy for most practical purposes. By combining light diffraction with forward light scattering, it is possible to increase the particle size range to the submicron region. In this way, one can measure the particle size distribution in the range 0.1–300 μm. Light diffraction gives an average diameter over all particle orientations as randomly oriented particles pass the light beam. A collimated and vertically polarized laser beam illuminates a particle dispersion and generates a diffraction pattern with the undiffracted beam in the center. The energy distribution of diffracted light is measured by a detector consisting of light-sensitive circles separated by isolating circles of equal width. The angle formed by the diffracted light increases with decreasing particle size. The angle-dependent intensity distribution is converted by Fourier optics into a spatial intensity distribution. The spatial intensity distribution is converted into a set of photocurrents and the particle size distribution is calculated using a computer. Several commercial instruments are available, e.g. Malvern MasterSizer (Malvern, UK), Horriba (Japan) and Coulter LS Sizer (USA).

Photon Correlation Spectroscopy (PCS) or Dynamic Light Scattering PCS is a technique that utilizes Brownian motion to measure the particle size. As a result of Brownian motion of dispersed particles, the intensity of scattered light undergoes fluctuations that are related to the velocity of the particles. Since larger particles move less rapidly than smaller particles, the intensity fluctuation (intensity versus time) pattern depends on particle size and this allows one to obtain the size distribution. In a system where the Brownian motion is not interrupted by sedimentation or particle–particle interaction, the movement of particles is random. Therefore, to apply the PCS technique, one must make sure that sedimentation does not occur during the measurement and the system is dilute enough to avoid particle–particle interactions.

The intensity fluctuation of the scattered light is measured using a photomultiplier and information on particle motion is obtained using a digital correlator. PCS allows one to measure the diffusion coefficient D of the particles, which is related to the particle radius R by the Stokes–Einstein equation:

$$D = \frac{kT}{6\pi\eta R} \tag{52}$$

where k is the Boltzmann constant, T is the absolute temperature and η is the viscosity of the medium.

The effect of particle interactions at a relatively low particle concentration c can be taken into account by expanding the diffusion coefficient into a power series of concentration:

$$D = D_0(1 + k_D c) \tag{53}$$

where D_0 is the diffusion coefficient at infinite dilution and k_D is the virial coefficient, which is related to particle interaction.

PCS is a rapid, absolute, non-destructive and rapid method for particle size measurements. It has some limitations; the main disadvantage is the poor resolution of the particle size distribution. Also, it suffers from the limited size range (absence of any sedimentation) that can be accurately measured. Several instruments are commercially available, e.g. from Malvern, Brookhaven and Coulter. The most recent instrument that is convenient to use is the HPPS, supplied by Malvern (UK), which allows one to measure the particle size distribution without the need for too much dilution (which may cause some particle dissolution).

Backscattering Technique This method is based on the use of fiber-optics, sometimes referred to as fiber-optic dynamic light scattering (FODLS), and it allows one to measure at high particle number concentrations. FODLS employs either one or two optical fibers. Alternatively, fiber bundles may be used. The exit port of the optical fiber (optode) is immersed in the sample and the scattered light in the same fiber is detected at a scattering angle of 180° (i.e. backscattering). This technique is suitable for on-line measurements during manufacture of a suspension. Several commercial instruments are available, e.g. from Lesentech (USA).

3.3 Wet Milling (Comminution)

The primary dispersion (sometimes referred to as the millbase) may be subjected to a bead milling process to produce nano-particles, which are essential for some coating applications. Subdivision of the primary particles into much smaller units in the nano-size range (10–100 nm) requires application of intense energy. In some cases, high-pressure homogenizers (such as the Microfluidizer, from Microfluidics, USA) may be sufficient to produce nano-particles. This is particularly the case with many organic pigments. In some cases, the high-pressure homogenizer is combined with application of ultrasound to produce nano-particles.

Milling or comminution (the generic term for size reduction) is a complex process and there is little fundamental information on its mechanism. For the breakdown of single crystals or particles into smaller units, mechanical energy is required. This energy in a bead mill is supplied by impaction of the glass or ceramic beads with the particles. As a result, permanent deformation of the particles and crack initiation result. This will eventually lead to the fracture of particles into smaller units. Since the milling conditions are random, some particles receive impacts far in excess of those required for fracture, whereas others receive impacts that are insufficient for the fracture process. This makes the milling operation grossly inefficient and only a small fraction of the applied energy is used in comminution. The rest of the energy is dissipated as heat, vibration, sound, interparticulate friction, etc.

The role of surfactants and dispersants in the grinding efficiency is far from being understood. In most cases, the choice of surfactants and dispersant is made by trial and error until a system is found that gives the maximum grinding efficiency. Rehbinder [26] investigated the role of surfactants in the grinding process. As a result of surfactant adsorption at the solid/liquid interface, the surface energy at the boundary is reduced, which facilitates the process of deformation or destruction. The adsorption of surfactants at the solid/liquid interface in cracks facilitates their propagation. This mechanism is referred to as the Rehbinder effect.

Several factors affect the efficiency of dispersion and milling: (i) the volume concentration of dispersed particles (i.e. the volume fraction); (ii) the nature of the wetting/dispersing agent; and (iii) the concentration of the wetter/dispersant (which determines the adsorption characteristics).

For optimization of the dispersion/milling process, the above parameters need to be systematically investigated. From the wetting performance of a surfactant, which can be evaluated using sinking time or contact angle measurements, one can establish the nature and concentration of the wetting agent. The nature and concentration of the dispersing agent required are determined by adsorption isotherm and rheological measurements (see Chapter 7).

Once the concentration of wetting/dispersing agent has been established, dispersions are prepared at various volume fractions keeping the ratio of the wetting/dispersing agent concentration to the solid content constant. Each system is then subjected to the dispersion/milling process keeping all parameters constant, namely (i) the speed of the stirrer (normally one starts at a lower speed and gradually increases it in increments at fixed time), (ii) the volume and size of the beads relative to the volume of the dispersion (an optimum value is required) and (iii) the speed of the mill.

The change of average particle size with time of grinding is established using, for example, a MasterSizer. Figure 3.26 shows a schematic representation of the reduction of particle size with grinding time using a typical bead mill (see below) at various volume fractions.

The presentation in Figure 3.26 is only schematic and is not based on experimental data: it shows the expected trend. When the volume fraction ϕ is below

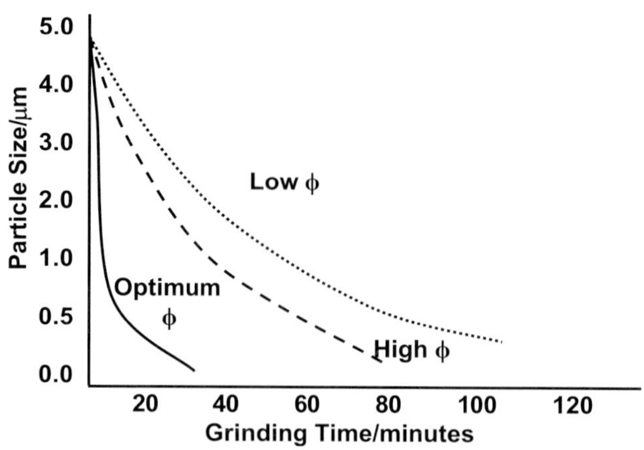

Figure 3.26 Variation of particle size with grinding time in a typical bead mill.

the optimum (in this case, the relative viscosity of the dispersion is low), a long time is required to achieve size reduction. In addition, the final particle size may be large and outside the nano-range. When ϕ is above the optimum value, the dispersion time is prolonged (due to the relatively high relative viscosity of the system) and the grinding time is also longer. In addition, the final particle size is larger than that obtained at the optimum ϕ. At the optimum volume fraction, both the dispersion and grinding time are shorter and also the final particle size is smaller.

3.3.1 Bead Mills

For the preparation of nano-dispersions, bead mills are commonly used. The beads are mostly made of glass or ceramics (which are preferred due to minimum contamination). The operating principle is to pump the premixed, preferably predispersed (using a high-speed mixer), millbase through a cylinder containing a specified volume of, say, ceramic beads (normally 0.5–1 mm in diameter to achieve nano-size particles). The dispersion is agitated by a single- or multi-disk rotor. The disk may be flat or perforated. The millbase passing through the shear zone is then separated from the beads by a suitable screen located at the opposite end of the feedport.

Generally, bead mills may be classified into two types: (i) vertical mills with an open or closed top and (ii) horizontal mills with closed chambers. The horizontal mills are more efficient and the most commonly used one are the Netzsch mill (Germany) and Dyno-Mill (Switzerland). These bead mills are available in various sizes from 0.5 to 500 l. The factors affecting the general dispersion efficiency are known reasonably well (from the manufacturers). The selection of the optimum diameter of the beads is important for maximum utilization. In general, the smaller the size of the beads and the higher their density, the more efficient the milling process is.

3.3.2 Principle of Operation of Bead Mills

The centrifugal force transmitted to the grinding beads at the tip of the rotating disk increases considerably with its weight, which applies greater shear to the mill base. This explains why denser beads are more efficient in grinding. The speed transmitted to the individual chambers of the beads at the tip of the disk assumes that the speed and the force can be calculated. The centrifugal force F is simply given by

$$F = \frac{v^2}{rg} \tag{54}$$

where v is the velocity, r is the radius of the disk and g is the acceleration due to gravity.

References

1 T.B. Blake, Wetting, in *Surfactants*, ed. Th.F. Tadros, Academic Press, London (1984).

2 G.D. Parfitt (ed.), *Dispersions of Powders in Liquids*, Applied Science, London (1977).

3 Th.F. Tadros (ed.), *Solid/Liquid Dispersions*, Academic Press, London (1987).

4 T. Young, *Philos. Trans. R. Soc. London*, *95*, 65 (1805).

5 J.W. Gibbs, *The Collected Work of J. Willard Gibbs*, Vol. 1, Longman Green, London (1928).

6 D.H. Everett, *Pure Appl. Chem.*, *52*, 1279 (1980).

7 R.E. Johnson, *J. Phys. Chem.*, *63*, 1655 (1959).

8 W.D. Harkins, *J. Phys. Chem.*, *5*, 135 (1937).

9 W.D. Harkins, *The Physical Chemistry of Surface Films*, Reinhold, New York (1952).

10 R.N. Wenzel, *Ind. Eng. Chem.*, *28*, 988 (1936).

11 A.B.D. Cassie and S. Dexter, *Trans. Faraday Soc.*, *40*, 546 (1944).

12 A.B.D. Cassie, *Discuss. Faraday Soc.*, *3*, 361 (1948).

13 W.A. Zisman, *Adv. Chem. Ser.*, *43*, 1 (1964).

14 C.A. Smolders, *Recl. Trav. Chim. Pays-Bas*, *80*, 650 (1960).

15 E.K. Rideal, *Philos. Mag.*, *44*, 1152 (1922).

16 E.D. Washburn, *Phys. Rev.*, *17*, 273 (1921).

17 S.S. Dukhin, G. Kretzscmar and R. Miller, *Dynamics of Adsorption at Liquid Interfaces*, Elsevier, Amsterdam (1995).

18 A.F.H. Ward and L. Tordai, *J. Phys. Chem.*, *14*, 453 (1946).

19 I. Panaitov and J.G. Petrov, *Ann. Univ. Sofia, Fac. Chem.*, *64*, 385 (1968/69).

20 R.S. Hansen, *J. Phys. Chem.*, *64*, 637 (1960).

21 R. Miller and K. Lunkenheimer, *Z. Phys. Chem.*, *259*, 863 (1978).

22 R. Zana, *Chem. Biol. Appl. Relaxation Spectrosc., Proc. NATO Adv. Study Inst., Ser. C*, *18*, 133 (1974).

23 R. Miller, A. Hoffmann, R. Hartmann, K.H. Schano and A. Halbig, *Adv. Mater.*, *4*, 370 (1992).

24 J.T. Davies and E.K. Rideal, *Interfacial Phenomena*, Academic Press, New York (1969).

25 J. Kloubek, K. Friml and F. Krejci, *Czech. Chem. Commun.*, *41*, 1845 (1976).

26 P.A. Rehbinder, *Colloid J. USSR*, *20*, 493 (1958).

4
Colloid Stabilization of Paint Dispersions

The colloid stabilization of the disperse phase in any paint formulation is essential during preparation of the pigment dispersion in order to prevent any particle reaggregation. The colloid stability is also very important for the long-term storage of a paint formulation. Aggregation of particles leads to poor performance of the paint, such as its opacity, color and durability. The colloid stability of any disperse system is determined by the balance between attractive (van der Waals) and repulsive forces. The latter can be of two types: (i) electrostatic double layer repulsion resulting from the presence of a charge on the particle surface; and (ii) steric repulsion arising from the presence of adsorbed nonionic surfactants or polymers. In some cases, a combination of electrostatic and steric repulsion (referred to as electrosteric) occurs, such as when using polyelectrolytes for stabilization. This chapter first describes the electrostatic double layer repulsion, van der Waals attraction and their combination, which lead to the general theory of colloid stability. The factors that affect the colloid stability, such as addition of electrolytes, are described in terms of the well-known theories of flocculation. This is followed by a section on the adsorption and conformation of polymeric surfactants at the solid/liquid interface. This is key to understanding how these polymeric surfactants can stabilize the particles against aggregation. Finally, the interaction between particles containing adsorbed layer is discussed and the factors responsible for steric stabilization are also considered.

4.1
Electrostatic Double Layer Repulsion

4.1.1
Structure of the Solid/Liquid Interface – Origin of Charges on Surfaces

A great variety of processes occur to produce a surface charge [1], as outlined below.

Surface ions are ions that have such a high affinity to the surface of the particles that they may be taken as part of the surface, e.g. Ag^+ and I^- for AgI.

Colloids and Interface Science Series, Vol. 6
Colloids in Paints. Tharwat F. Tadros

ISBN: 978-3-527-31466-9

For AgI in a solution of KNO_3, the surface charge σ_0 is given by the following expression:

$$\sigma_p = F(\Gamma_{Ag^+} - \Gamma_{I^-}) = F\Gamma_{AgNO_3} - \Gamma_{KI} \quad (1)$$

where F is the Faraday constant (96 500 C mol^{-1}) and G is the surface excess of ions (mol m^{-2}).

Similarly for an oxide such as silica, titania or alumina in KNO_3, H^+ and OH^- may be taken as part of the surface:

$$\sigma_0 = F(\Gamma_{H^+} - \Gamma_{OH^-}) = F(\Gamma_{HCl} - \Gamma_{KOH}) \quad (2)$$

The ions which determine the charge on the surface are termed potential-determining ions.

Consider an oxide surface as illustrated in Figure 4.1. The charge depends on the pH of the solution: below a certain pH the surface is positive and above a certain pH it is negative. At a specific pH ($\Gamma_H = \Gamma_{OH}$) the surface is uncharged; this is referred to as the point of zero charge (pzc).

The pzc depends on the type of the oxide: for an acidic oxide such as silica the pzc is ~pH 2–3 and for a basic oxide such as alumina it is ~pH 9. For an amphoteric oxide such as titania, the pzc ~pH 6.

In some cases, specifically adsorbed ions (that have non-electrostatic affinity to the surface) 'enrich' the surface but may not be considered as part of the surface, e.g. bivalent cations on oxides and cationic and anionic surfactants on most surfaces [1].

Isomorphic substitution [2] can occur, e.g. with sodium montmorillonite, i.e. replacement of cations inside the crystal structure by cations of lower valency, e.g. Si^{4+} replaced by Al^{3+}. The deficit of one positive charge gives one negative charge. The surface of Na montmorillonite is negatively charged with Na^+ as counterions. This is illustrated schematically in Figure 4.2. The surface charge + counterions form the electrical double layer.

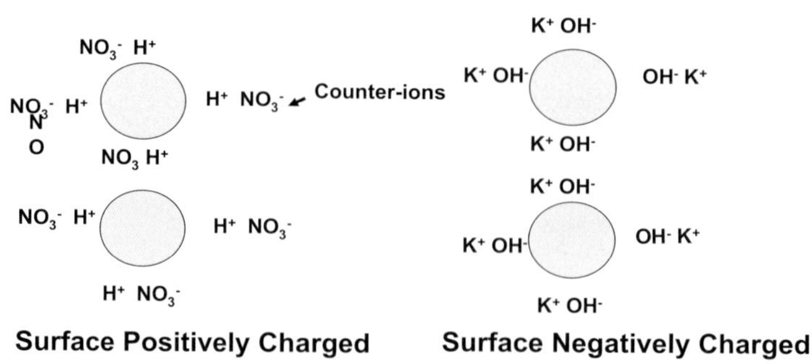

Figure 4.1 Schematic representation of an oxide surface.

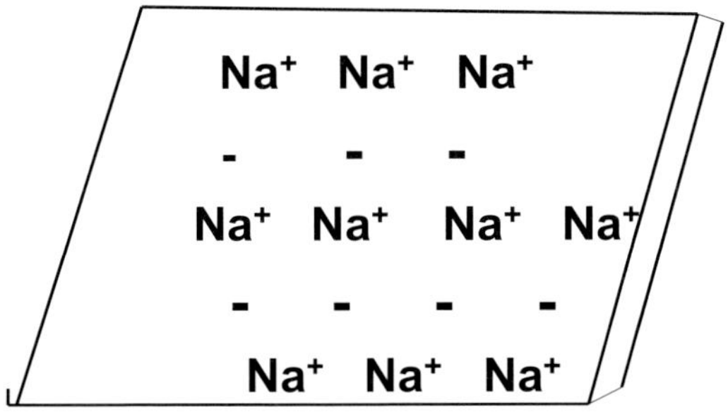

Figure 4.2 Schematic representation of a clay particle.

4.1.2
Structure of the Electrical Double Layer

4.1.2.1 Diffuse Double layer (Gouy and Chapman)

The surface charge σ_0 is compensated by unequal distribution of counterions (opposite in charge to the surface) and co-ions (same sign as the surface) which extend to some distance from the surface [3, 4]. This is represented schematically in Figure 4.3.

The potential decays exponentially with distance x. At low potentials

$$\psi = \psi_0 \exp(-\kappa x) \tag{3}$$

Note that when $x = 1/\kappa$, $\psi_x = \psi_0/e$; $1/\kappa$ is referred to as the 'thickness of the double layer'.

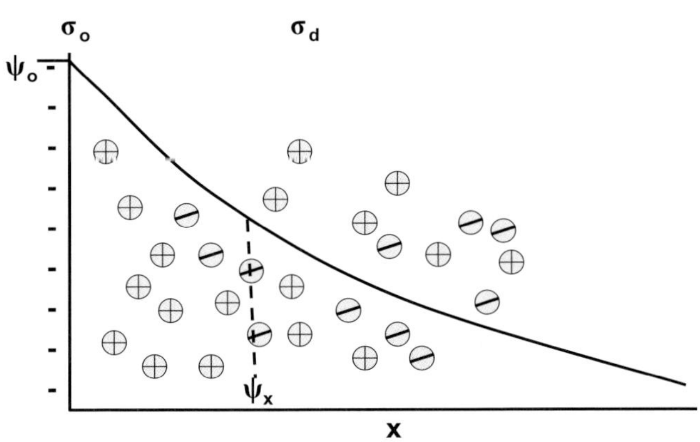

Figure 4.3 Schematic representation of the diffuse double layer according to Gouy and Chapman.

Table 4.1 Values of $1/\kappa$ for 1:1 electrolyte at 25 °C.

C/mol dm^{-3}	$(1/\kappa)$/nm
10^{-5}	100
10^{-4}	33
10^{-3}	10
10^{-2}	3.3
10^{-1}	1

The double layer extension depends on the electrolyte concentration and valency of the counterions:

$$\frac{1}{\kappa}=\left(\frac{\varepsilon_r\varepsilon_0 kT}{2n_0 Z_i{}^2 e^2}\right)^{\frac{1}{2}} \tag{4}$$

where ε_r is the permittivity (dielectric constant) and is 78.6 for water at 25 °C, ε_0 is the permittivity of free space, k is the Boltzmann constant, T is the absolute temperature, n_0 is the number of ions per unit volume of each type present in bulk solution, Z_i is the valency of the ions and e is the electronic charge. Values of $1/\kappa$ for a 1:1 electrolyte (e.g. KCl) are given in Table 4.1. The double layer extension increases with decrease in electrolyte concentration.

4.1.2.2 **Stern–Grahame Model of the Double Layer**

Stern [5] introduced the concept of the non-diffuse part of the double layer for specifically adsorbed ions, the rest being diffuse in nature. This is illustrated schematically in Figure 4.4.

The potential decreases linearly in the Stern region and then exponentially. Grahame distinguished two types of ions in the Stern plane, physically adsorbed counterions (outer Helmholtz plane) and chemically adsorbed ions (that lose part of their hydration shell) (inner Helmholtz plane).

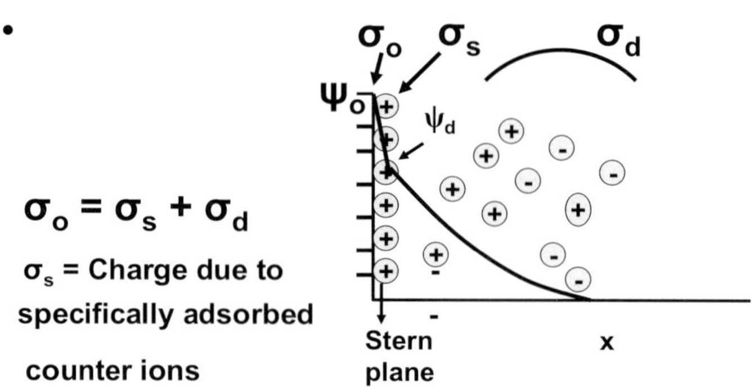

Figure 4.4 Schematic representation of the double layer according to Stern and Grahame.

4.1.3
Electrical Double Layer Repulsion

When charged colloidal particles in a dispersion approach each other such that the double layers begin to overlap (the particle separation becomes less than twice the double layer extension), repulsion occurs. The individual double layers can no longer develop unrestrictedly, since the limited space does not allow complete potential decay [6]. This is illustrated in Figure 4.5 for two flat plates. The potential $\psi_{h/2}$ half way between the plates is no longer zero (as would be the case for isolated particles at $x \to \infty$).

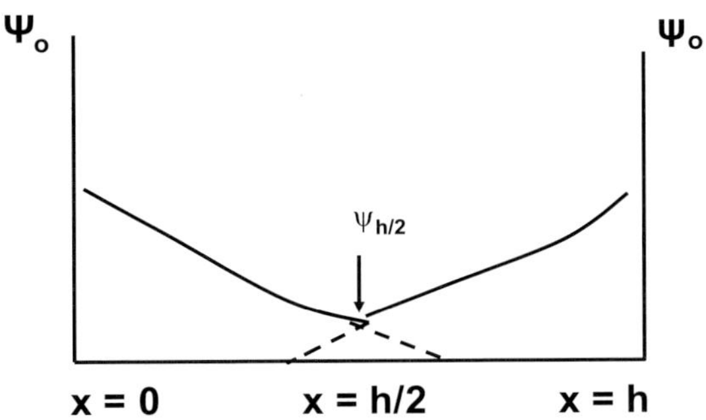

Figure 4.5 Schematic representation of double layer interaction for two flat plates.

For two spherical particles of radius R and surface potential ψ_0 and condition $\kappa R < 3$, the expression for the electrical double layer repulsive interaction is

$$G_{\text{el}} = \frac{4\pi\varepsilon_r\varepsilon_0 R^2 {\psi_0}^2 \exp(-\kappa h)}{2R + h} \tag{5}$$

where h is the closest distance of separation between the surfaces. This expression shows the exponential decay of G_{el} with h. The higher the value of κ (i.e. the higher the electrolyte concentration), the steeper is the decay, as shown schematically in Figure 4.6. This means that at any given distance h, the double layer repulsion decreases with increase in electrolyte concentration.

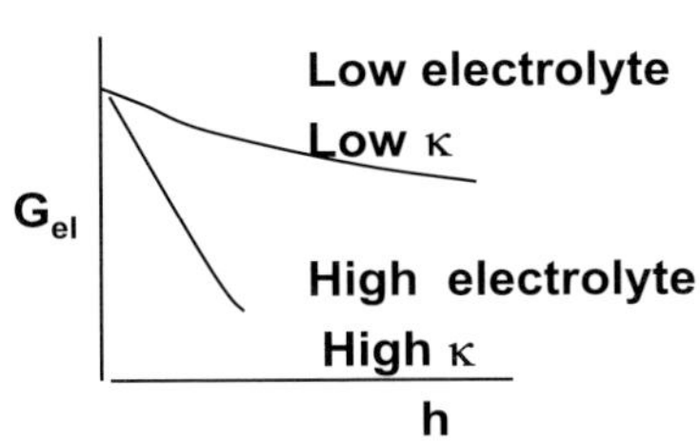

Figure 4.6 Variation of G_{el} with h at different electrolyte concentrations.

4.1.4
Van der Waals Attraction

As is well known, atoms or molecules always attract each other at short distances of separation. The attractive forces are of three different types: dipole–dipole interaction (Keesom), dipole–induced dipole interaction (Debye) and London dispersion force. The London dispersion force is the most important, since it occurs for polar and non-polar molecules. It arises from fluctuations in the electron density distribution.

At small distances of separation r in vacuum, the attractive energy between two atoms or molecules is given by

$$G_{\mathrm{aa}} = -\frac{\beta_{11}}{r^6} \tag{6}$$

where β_{11} is the London dispersion constant.

For colloidal particles which are made of atom or molecular assemblies, the attractive energies may be added and this results in the following expression for two spheres (at small h) [7]:

$$G_{\mathrm{A}} = -\frac{A_{11(2)}R}{12h} \tag{7}$$

where $A_{11(2)}$ is the effective Hamaker constant of two identical particles with Hamaker constant A_{11} in a medium with Hamaker constant A_{22}. When the particles are dispersed in a liquid medium, the van der Waals attraction has to be modified to take into account the medium effect. When two particles are brought from infinite distance to h in a medium, an equivalent amount of medium has to be transported the other way round. Hamaker forces in a medium are excess forces.

Consider two identical spheres 1 at a large distance apart in a medium 2 as illustrated in Figure 4.7a. In this case, the attractive energy is zero [8]. Figure

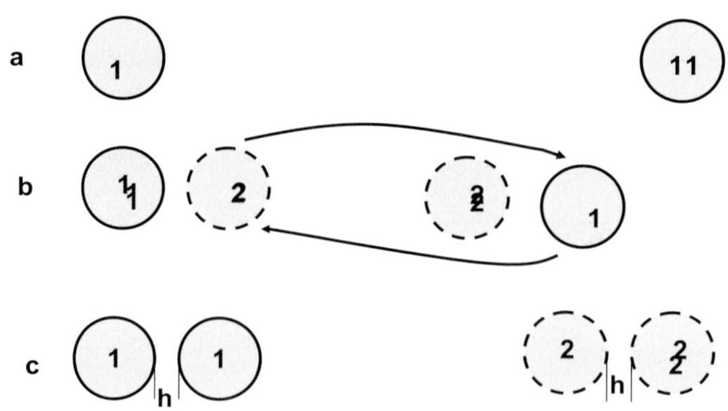

Figure 4.7 Schematic representation of the interaction of two particles in a medium.

Table 4.2 Hamaker constants for some liquids.

Liquid	$A_{22} \times 10^{20}$/J
Water	3.7
Ethanol	4.2
Decane	4.8
Hexadecane	5.2
Cyclohexane	5.2

4.7b gives the same situation with arrows indicating the exchange of 1 against 2. Figure 4.7c shows the complete exchange, which now shows the attraction between the two particles 1 and 1 and equivalent volumes of the medium 2 and 2.

The effective Hamaker constant for two identical particles 1 and 1 in a medium 2 is given by

$$A_{11(2)} = A_{11} + A_{22} - 2A_{12} = (A_{11}^{\frac{1}{2}} - A_{22}^{\frac{1}{2}})^2 \quad (8)$$

Equation 8 shows that two particles of the same material attract each other unless their Hamaker constants exactly match each other. The Hamaker constant of any material is given by

$$A = \pi q^2 \beta_{ii} \quad (9)$$

where q is the number of atoms or molecules per unit volume.

In most cases, the Hamaker constant of the particles is higher than that of the medium. Examples of Hamaker constants for some liquids are given in Table 4.2 and Table 4.3 gives values of the effective Hamaker constant for some particles in some liquids. Generally, the effect of the liquid medium is to reduce the Hamaker constant of the particles below its value in vacuum (air).

Table 4.3 Effective Hamaker constants of some particles in water.

System	$A_{11(2)} \times 10^{20}$/J
Fused quartz–water	0.83
Al_2O_3–water	5.32
Copper–water	30.00
Poly(methyl methacrylate)–water	1.05
Poly(vinyl chloride)–water	1.03
Polytetrafluoroethylene–water	0.33

G_A decreases with increase in h, as shown schematically in Figure 4.8. At very short distances, Born repulsion appears.

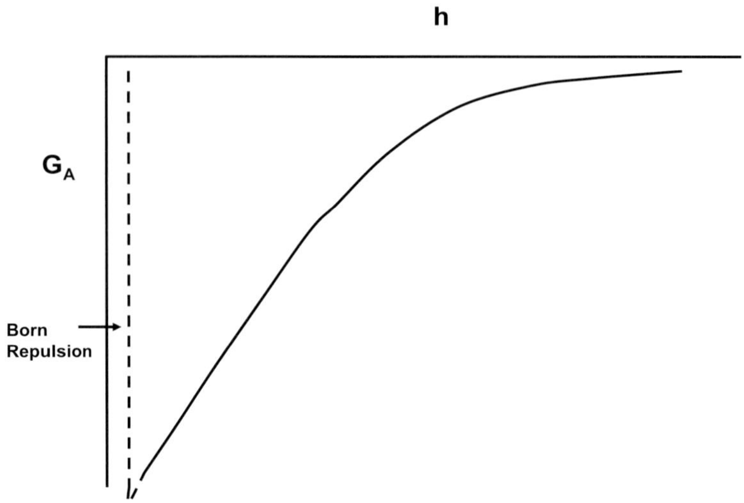

Figure 4.8 Variation of G_A with h.

4.2
Total Energy of Interaction

4.2.1
Deryaguin–Landau–Verwey–Overbeek (DLVO) Theory [9, 10]

Combination of G_{el} and G_A results in the well-known theory of stability of colloids (DLVO theory) [9, 10]:

$$G_T = G_{el} + G_A \tag{10}$$

A plot of G_T versus h is shown in Figure 4.9, which represents the case at low electrolyte concentrations, i.e. strong electrostatic repulsion between the particles. G_{el} decays exponentially with h, i.e. $G_{el} \to 0$ as h becomes large. $G_A \propto 1/h$, i.e. G_A does not decay to 0 at large h.

At long distances of separation, $G_A > G_{el}$, resulting in a shallow minimum (secondary minimum). At very short distances, $G_A \gg G_{el}$, resulting in a deep primary minimum. At intermediate distances, $G_{el} > G_A$, resulting in energy maximum, G_{max}, the height of which depends on ψ_0 (or ψ_d) and the electrolyte concentration and valency.

At low electrolyte concentrations ($<10^{-2}$ mol dm^{-3} for a 1:1 electrolyte), G_{max} is high ($>25kT$), which prevents particle aggregation into the primary minimum.

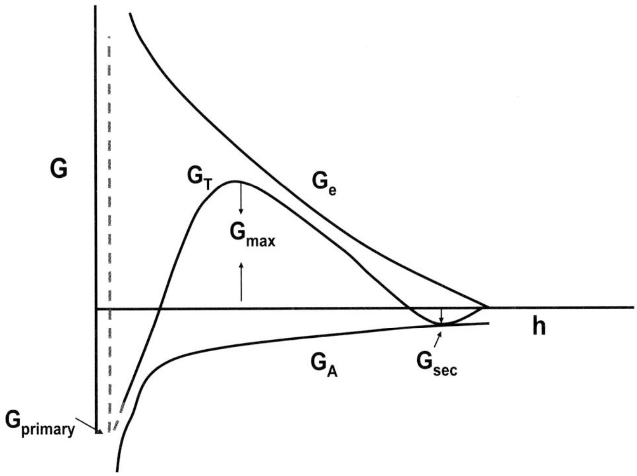

Figure 4.9 Schematic representation of the variation of G_T with h according to the DLVO theory.

The higher the electrolyte concentration (and the higher the valency of the ions), the lower is the energy maximum.

Under some conditions (depending on electrolyte concentration and particle size), flocculation into the secondary minimum may occur. This flocculation is weak and reversible. By increasing the electrolyte concentration, G_{max} decreases until at a given concentration it vanishes and particle coagulation occurs. This is illustrated in Figure 4.10, which shows the variation of G_T with h at various electrolyte concentrations.

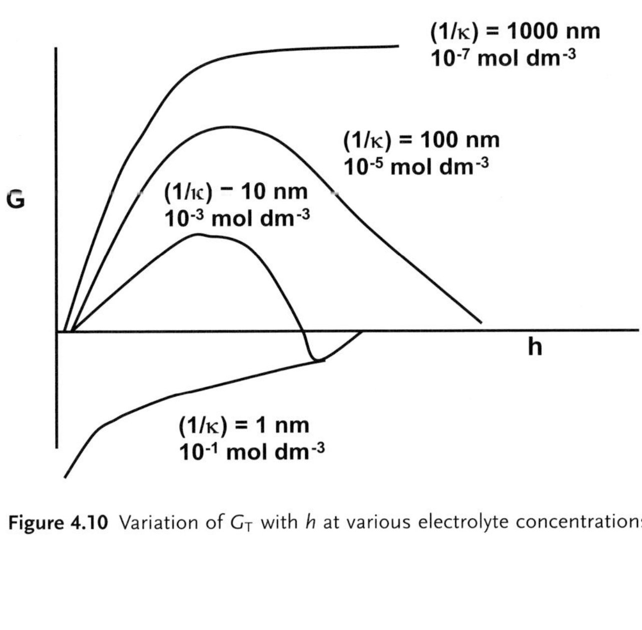

Figure 4.10 Variation of G_T with h at various electrolyte concentrations.

4.2.2
Flocculation of Electrostatically Stabilized Suspensions

As discussed earlier, the condition for kinetic stability is $G_{max} > 25kT$. When $G_{max} < 5kT$, flocculation occurs. Two types of flocculation kinetics may be distinguished: fast flocculation with no energy barrier and slow flocculation when an energy barrier exists.

The fast flocculation kinetics were treated by von Smoluchowski [11], who considered the process to be represented by second-order kinetics and the process is simply diffusion controlled. The number of particles n at any time t may be related to the initial number (at $t = 0$) n_0 by the following expression:

$$n = \frac{n_0}{1 + kn_0t} \tag{11}$$

where k is the rate constant for fast flocculation that is related to the diffusion coefficient of the particles D, i.e.

$$k = 8\pi DR \tag{12}$$

D is given by the Stokes–Einstein equation:

$$D = \frac{kT}{6\pi\eta R} \tag{13}$$

Combining equations 12 and 13:

$$k = \frac{4}{3}\frac{kT}{\eta} = 5.5 \times 10^{-18}\ \mathrm{m^3\ s^{-1}}\ \text{for water at 25 °C} \tag{14}$$

The half-life $t_{\frac{1}{2}}$ $(n = \frac{1}{2}n_0)$ can be calculated at various n_0 or volume fraction ϕ as give in Table 4.4.

The slow flocculation kinetics were treated by Fuchs [12], who related the rate constant k to the Smoluchowski rate by the stability constant W:

Table 4.4 Half-life of dispersion flocculation.

R/μm	$\phi = 10^{-5}$	$\phi = 10^{-2}$	$\phi = 10^{-1}$	$\phi = 5 \times 10^{-1}$
0.1	765 s	76 ms	7.6 ms	1.5 ms
1.0	21 h	76 s	7.6 s	1.5 s
10.0	4 months	21 h	2 h	25 m

$$W = \frac{k_0}{k} \tag{15}$$

W is related to $G_{\max}$ by the following expression [13]:

$$W = \frac{1}{2} k \exp\left(\frac{G_{\max}}{kT}\right) \tag{16}$$

Since $G_{\max}$ is determined by the salt concentration C and valency, one can derive an expression relating W to C and Z [13]:

$$\log W = -2.06 \times 10^9 \left(\frac{R\gamma^2}{Z^2}\right) \log C \tag{17}$$

where γ is a function that is determined by the surface potential ψ_0:

$$\gamma = \frac{\exp(Ze\psi_0/kT) - 1}{\exp(Ze\psi_0/kT) + 1} \tag{18}$$

Plots of log W versus log C are shown in Figure 4.11. The condition log $W = 0$ ($W = 1$) is the onset of fast flocculation. The electrolyte concentration at this point defines the critical flocculation concentration (CFC). Above the CFC, $W < 1$ (due to the contribution of van der Waals attraction, which accelerates the rate above the Smoluchowski value). Below the CFC, $W > 1$, and it increases with decrease in electrolyte concentration. Figure 4.11 also shows that the CFC decreases with increase in valency, in accordance with the Scultze–Hardy rule.

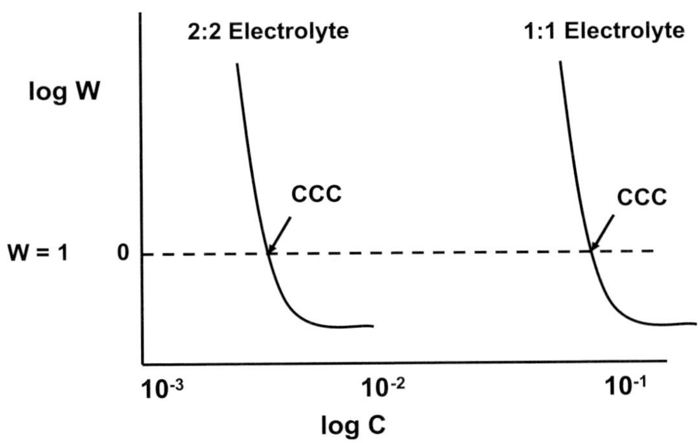

Figure 4.11 Log W–log C curves for electrostatically stabilized dispersions.

Another mechanism of flocculation is that involving the secondary minimum (G_{min}), which is a few kT units. In this case, flocculation is weak and reversible, hence one must consider both the rate of flocculation (forward rate k_f) and deflocculation (backward rate k_b). In this case, the rate of or decrease in particle number with time is given by the expression

$$-\frac{dn}{dt} = -k_f n^2 + k_b n \tag{19}$$

The backward reaction (break-up of weak flocs) reduces the overall rate of flocculation.

4.2.3
Criteria for Stabilization of Dispersions with Double Layer Interaction

The two main criteria for stabilization are: (i) high surface or Stern potential (zeta potential) and high surface charge, and (ii) low electrolyte concentration and low valency of counter- and co-ions. One should ensure that an energy maximum in excess of $25kT$ exists on the energy–distance curve. When $G_{max} \gg kT$, the particles in the dispersion cannot overcome the energy barrier, thus preventing coagulation.

In some cases, particularly with large and asymmetric particles, flocculation into the secondary minimum may occur. This flocculation is usually weak and reversible and may be advantageous for preventing the formation of hard sediments.

4.2.4
Electrokinetic Phenomena and Zeta Potential

As mentioned above, one of the main criteria for electrostatic stability is the high surface or zeta potential, which can be measured experimentally, as discussed below. Before describing the experimental techniques for measuring the zeta potential, it is essential to consider the electrokinetic effects in some detail, describing the theories that can be used to calculate the zeta potential form the particle electrophoretic mobility [14].

Electrokinetic effects are the direct result of charge separation at the interface between two phases. Consider a negatively charged surface; positive ions (counterions) are attracted to the surface, whereas negative ions (co-ions) are repelled. This is shown schematically in Figure 4.12.

The accumulation of excess positive ions causes a gradual reduction in the potential from its value ψ_0 at the surface to zero in bulk solution. At a point p from the surface, one can define a potential ψ_x.

Electrokinetic effects arise when one of the two phases is caused to move tangentially past the second phase. Tangential motion can be caused by an electric

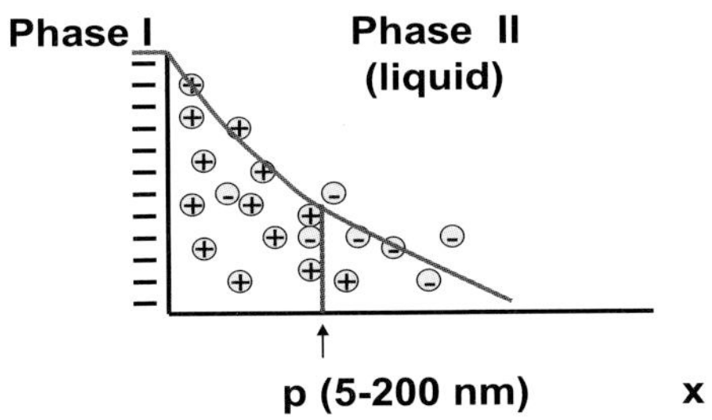

Figure 4.12 Schematic representation of charge accumulation at an interface.

field, forcing a liquid in a capillary, or a gravitational field on the particles, which leads to four different types of electrokinetic phenomena:

1. *Electrophoresis:* Movement of one phase induced by application of an external electric field. One measures the particle velocity v from which the electrophoretic mobility u can be calculated:

$$u = \frac{v}{E/l} \text{ m}^2 \text{ V}^{-1} \text{ s}^{-1} \tag{20}$$

   where E is the applied potential and l is the distance between the two electrodes; E/l is the field strength.
2. *Electroosmosis:* The solid is kept stationary (e.g. in the form of a glass tube) and the liquid is allowed to move under the influence of an electric field. The applied field acts on the charges (ions in the liquid) and when these move they drag liquid with them.
3. *Streaming potential:* The liquid is forced through a capillary or a porous plug (containing the particles) under the influence of a pressure gradient. The excess charges near the wall (or the surface of particles in the plug) are carried along by the liquid flow, thus producing an electric field that can be measured by using electrodes and an electrometer.
4. *Sedimentation potential (Dorn effect):* The particles are allowed to settle or rise through a fluid under the influence of gravity (or using a centrifugal force). When the particles move, they leave behind their ionic atmosphere, which creates a potential difference in the direction of motion that can be measured using electrodes and an electrometer.

In this section, only electrophoresis will be discussed, since this is the most commonly used method for dispersions, allowing one to measure the particle mobility, which can be converted to the zeta potential using theoretical treatments.

In all electrokinetic phenomena [14], a fluid moves with respect to a solid surface. One needs to derive a relationship between fluid velocity (which varies with distance from the solid) and the electric field in the interfacial region.

The most important concept is the surface of shear, an imaginary surface close to the surface within which the fluid is stationary. This is illustrated in Figure

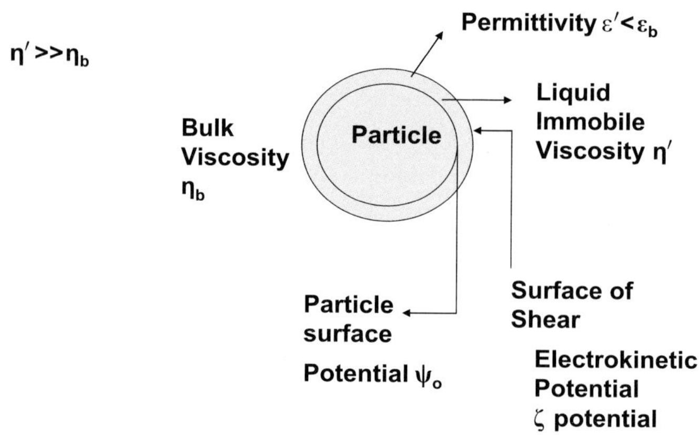

Figure 4.13 Schematic representation of the shear plane.

4.13, which shows the position of the surface potential ψ_0, the shear plane and zeta potential (which is close to the Stern potential ψ_d).

Measurement of the zeta potential (ζ) is valuable in determining the properties of dispersions. In addition, it has many other applications in various fields, such as electrode kinetics, electrodialysis, corrosion, adsorption of surfactants and polymers, crystal growth, mineral flotation and particle sedimentation.

Although the measurement of particle mobility is fairly simple (particularly with the development of automated instruments), the interpretation of the results is not simple. The calculation of zeta potential from particle mobility is not straightforward since it depends on the particle size and shape and also the electrolyte concentration. For simplicity, we will assume that the particles are spherical.

4.2.5 Calculation of Zeta Potential

4.2.5.1 Von Smoluchowski (Classical) Treatment [15]

This applies to the case where the particle radius R is much larger than the double layer thickness $(1/\kappa)$, i.e. $\kappa R \gg 1$. This generally applies to particles that are larger than 0.5 μm (when the 1:1 electrolyte concentration is lower than 10^{-3} mol dm^{-3}, i.e. $\kappa R > 10$):

$$u = \frac{\varepsilon_r \varepsilon_0 \zeta}{\eta} \tag{21}$$

where ε_r is the relative permittivity of the medium, 78.6 for water at 25 °C, ε_0 is the permittivity of free space, 8.85×10^{-12} F m^{-1}, η is the viscosity of the medium, 8.9×10^{-4} Pa s for water at 25 °C, and ζ is the zeta potential in volts.

For water at 25 °C,

$$\zeta = 1.282 \times 10^6 u \tag{22}$$

where u is expressed in $m^2\ V^{-1}\ s^{-1}$.

4.2.5.2 The Hückel Equation [16]

This applies for the case where $\kappa R < 1$:

$$u = \frac{2}{3}\frac{\varepsilon_r \varepsilon_0 \zeta}{\eta} \tag{23}$$

This equation applies for small particles (<100 nm) and thick double layers (low electrolyte concentration).

4.2.5.3 Henry's Treatment [17]

Henry accounted for the discrepancy between von Smoluchowski's and Hückel's treatments. Hückel disregarded the deformation of the electric field by the particle, whereas Smoluchowski assumed the field to be uniform and everywhere parallel to the particle surface. These two assumptions are justified in the extreme cases of $\kappa R \ll 1$ and $\kappa R \gg 1$, respectively.

For the intermediate cases where κR is not too small or too large, Henry derived the following expression (which can be applied at all κR values):

$$u = \frac{\varepsilon_r \varepsilon_0 \zeta}{\eta} f(\kappa R) \tag{24}$$

The function $f(\kappa R)$ depends also on the particle shape. Values of $f(\kappa R)$ at various values of κR are given in Table 4.5.

Henry's calculations are based on the assumption that the external field can be superimposed on the field due to the particle and hence it can only be applied for low potentials ($\zeta < 25$ mV). It also does not take into account the distortion of the field induced by the movement of the particle (relaxation effect).

Table 4.5 Henry's correction factor $f(\kappa R)$.

κR	$f(\kappa R)$	κR	$f(\kappa R)$
0	1.0	5	1.160
1	1.027	10	1.239
2	1.066	25	1.370
3	1.101	100	1.460
4	1.133	∞	1.500

Wiersema *et al.* [18] introduced two corrections for Henry's treatment, namely the relaxation and retardation (movement of the liquid with the double layer ions) effects, and a numerical tabulation of the relation between mobility and zeta potential was given by Ottewill and Shaw [19]. Such tables are useful for conversion of u to ζ at all practical values of κR.

4.2.6 Measurement of Electrophoretic Mobility

4.2.6.1 Ultramicroscopic Technique (Microelectrophoresis)

This is the most commonly used method since it allows direct observation of the particles using an ultramicroscope (suitable for particles that are larger than 100 nm). Basically, a dilute suspension is placed in a cell consisting of a thin-walled (~100 μm) glass tube that is attached to two larger bore tubes with sockets for placing the electrodes. The cell is immersed in a thermostated bath (accurate to ±0.1 °C) that contains an attachment for illumination and a microscope objective for observing the particles. It is also possible to use a video camera for directly observing the particles.

Since the glass walls are charged (usually negative at practical pH measurements), the solution in the cell will in general experience electro-osmotic flow. Only where the electro-osmotic flow is zero, i.e. at the stationary level, can the electrophoretic mobility of the particles be measured. The stationary level is located at a distance of 0.707 times the radius from the center of the tube or 0.146 times the internal diameter from the wall.

By focusing the microscope objective at the top and bottom of the walls of the tube, one can easily locate the position of the stationary levels. The average particle velocity is measured at the top and bottom stationary levels by averaging at least 20 measurements in each direction (the eyepiece of the microscope is fitted with a graticule).

Several commercial instruments are available (e.g. from Rank Brothers, Bottisham, Cambridge, UK, and Pen Kem, Minneapolis, MN, USA).

For large particles (>1 μm and high density), sedimentation may occur during the measurement. In this case, one can use a rectangular cell and observe the particles horizontally from the side of the glass cell.

Microelectrophoresis has many advantages since the particles can be measured in their normal environment. It is preferable to dilute the suspension with the supernatant liquid which can be produced by centrifugation.

4.2.6.2 Laser Velocimetry Technique

This method is suitable for small particles that undergo Brownian motion. The light scattered by small particles will show intensity fluctuations as a result of Brownian diffusion (Doppler shift). By application of an electric field as the particles undergo Brownian motion and measuring the fluctuation in intensity of the scattered light (using a correlator), one can measure the particle mobility.

Two laser beams of equal intensity are allowed to cross at a particular point within the cell containing the suspension of particles. At the intersection of the beam, which is focused at the stationary level, interferences of known spacing are formed. The particles moving through the fringes under the influence of the electric field scatter light whose intensity fluctuates with a frequency that is related to the mobility of the particles.

The photons are detected by a photomultiplier and the signal is fed to the correlator. The resulting correlation function is analyzed to determine the frequency (Doppler) spectrum and this is converted to the particle velocity v:

$$v = \Delta \nu s \tag{25}$$

where Δv is the Doppler shift frequency and s is the spacing between the interference fringes in the region where the beams cross; s is given by the relationship

$$s = \frac{\lambda}{2 \sin\left(\frac{\alpha}{2}\right)} \tag{26}$$

where λ is the laser wavelength and α is the angle between the crossing laser beams.

The velocity spectrum is then converted to a mobility spectrum (allowing one to obtain the mobility distribution) and the mobility is converted to zeta potential using Hückel's equation.

Several commercial instruments are available, e.g. the Malvern Zeta Sizer and Coulter Delsa Sizer.

4.3 Adsorption and Conformation of Polymeric Surfactants at Interfaces

Understanding the adsorption and conformation of polymeric surfactants at interfaces is key to knowing how these molecules act as stabilizers. Most basic ideas on adsorption and conformation of polymers have been developed for the solid/liquid interface [20].

The process of polymer adsorption is fairly complicated; it involves polymer/surface interaction, polymer/solvent interaction, surface/solvent interaction and the configuration (conformation) of the polymer at the solid/liquid interface. The polymer/surface interaction is described in terms of adsorption energy per segment, χ^s. The polymer/solvent interaction is described in terms of the Flory–Huggins interaction parameter, χ. The polymer configuration is described by sequences of: trains, segments in direct contact with the surface; loops, segments between the trains that extend into solution; and tails, ends of the molecules that also extend into solution. A schematic representation of the various polymer configurations is given in Figure 4.14.

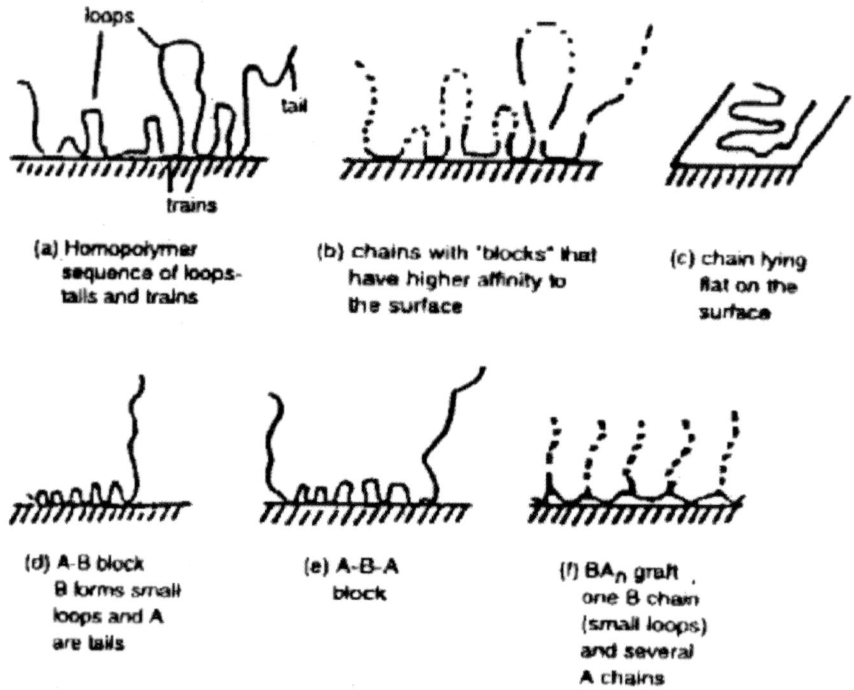

Figure 4.14 Various conformations of macromolecules on a plane surface.

For homopolymers, e.g. poly(ethylene oxide) (PEO) and polyvinylpyrrolidone (PVP), a train–loop–tail configuration is the case. For adsorption to occur, a minimum energy of adsorption per segment χ^s is required. When a polymer molecule adsorbs on a surface, it loses configurational entropy and this must be compensated for by an adsorption energy χ^s per segment. This is shown schematically in Figure 4.15, where the amount adsorbed Γ is plotted versus χ^s The minimum value of χ^s can be very small ($<0.1kT$), since a large number of segments per molecule are adsorbed. For a polymer with say 100 segments, 10% of which are in trains, the adsorption energy per molecule now reaches $1kT$ (with $\chi^s = 0.1kT$). For 1000 segments, the adsorption energy per molecule is now $10kT$.

Homopolymers are not the most suitable for stabilization of dispersions. For strong adsorption, one needs the molecule to be 'insoluble' in the medium and to show strong affinity ('anchoring') to the surface. For stabilization, one needs the molecule to be highly soluble in the medium and strongly solvated by its molecules – this requires a Flory–Huggins interaction parameter of <0.5. The above opposing effects can be resolved by introducing 'short' blocks in the molecule which are insoluble in the medium and have a strong affinity to the surface, as illustrated in Figure 4.14b. An example is partially hydrolyzed poly(vinyl acetate) (88% hydrolyzed, i.e. with 12% acetate groups), usually referred to as poly(vinyl alcohol) (PVA):

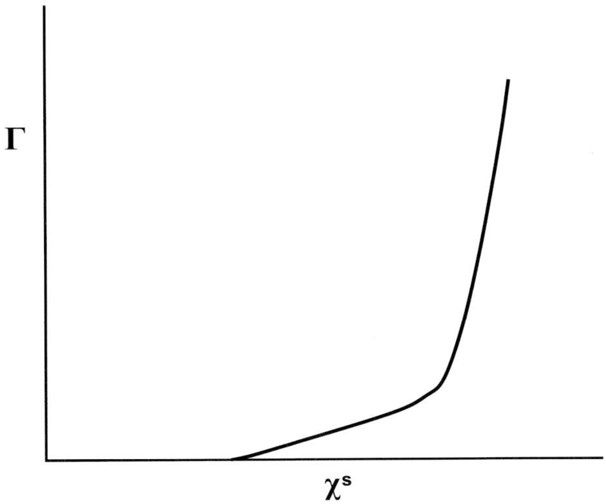

Figure 4.15 Variation of amount adsorbed Γ with adsorption energy per segment χ^s.

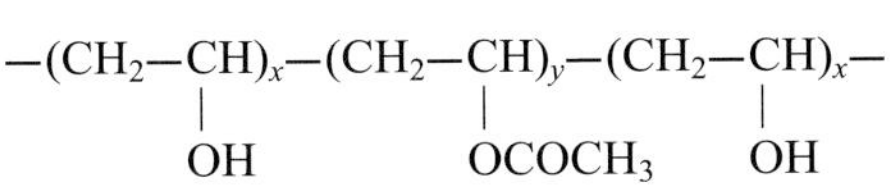

Structure 4.1

The above requirements are better satisfied using A–B, A–B–A and BA_n graft copolymers. B is chosen to be highly insoluble in the medium and it should have high affinity to the surface. This is essential to ensure strong 'anchoring' to the surface (irreversible adsorption). A is chosen to be highly soluble in the medium and strongly solvated by its molecules. The Flory–Huggins parameter χ can be applied in this case. For a polymer in a good solvent, χ has to be lower than 0.5; the smaller the χ value, the better is the solvent for the polymer chains. Examples of B for hydrophobic particles in aqueous media are polystyrene (PS) and poly(methyl methacrylate). Examples of A in aqueous media are poly(ethylene oxide), poly(acrylic acid), polyvinylpyrrolidone and polysaccharides. For non-aqueous media such as hydrocarbons, the A chain(s) could be poly(12-hydroxystearic acid).

For a full description of polymer adsorption, one needs to obtain information on the following:

1. The amount of polymer adsorbed Γ (in milligrams or moles) per unit area of the particles. It is essential to know the surface area of the particles in the suspension. Nitrogen adsorption on the powder surface may give such information (by application of the BET equation) provided that there will be no change in area on dispersing the particles in the medium. For many practical systems,

a change in surface area may occur on dispersing the powder, in which case one has to use dye adsorption to measure the surface area (some assumptions have to be made in this case).

2. The fraction of segments in direct contact with the surface, i.e. the fraction of segments in trains p ($p =$ number of segments in direct contact with the surface/total number).
3. The distribution of segments in loops and tails, $\rho(z)$, which extend in several layers from the surface. $\rho(z)$ is usually difficult to obtain experimentally, although recently application of small angle neutron scattering has been used to obtain such information. An alternative and useful parameter for assessing 'steric stabilization' is the hydrodynamic thickness, δ_h (thickness of the adsorbed or grafted polymer layer plus any contribution from the hydration layer). Several methods can be applied to measure δ_h, as will be discussed below.

4.3.1 Theories of Polymer Adsorption

Two main approaches have been developed to treat the problem of polymer adsorption: (i) the random walk approach [20], which is based on Flory's treatment of the polymer chain in solution; the surface being considered as a reflecting barrier; and (ii) the statistical mechanical approach [21, 22], where the polymer configuration was treated as being made of three types of structures, trains, loops and tails, each having a different energy state.

The random walk approach is an unrealistic model for the problem of polymer adsorption, since the polymer interacts in a specific manner with the surface and the solvent. The statistical mechanical approach is a more realistic model for the problem of polymer adsorption since it takes into account the various interactions involved. A useful model for treating polymer adsorption and configuration was suggested by Scheutjens and Fleer (SF theory) [23–27], which is referred to as the step-weighted random walk approach and is summarized below.

The SF theory starts without any assumption about the segment density distribution. The partition functions were derived for the mixture of free and adsorbed polymer molecules, and also for the solvent molecules. All chain conformations were described as step-weighted random walks on a quasi-crystalline lattice which extends in parallel layers from the surface – this is shown schematically in Figure 4.16. The partition function is written in terms of a number of configurations. These were treated as connected sequences of segments. In each layer, random mixing between segments and solvent molecules was assumed (mean field approximation). Each step in the random walk was assigned a weighting factor p_i.

The weighting factor p_i consists of three contributions: (i) an adsorption energy χ^s (which exists only for the segments that are near the surface); (ii) the configurational entropy of mixing (which exists in each layer); and (iii) the segment–solvent interaction parameter χ (the Flory–Huggins interaction parameter; note that $\chi = 0$ for an athermal solvent and $\chi = 0.5$ for a θ-solvent).

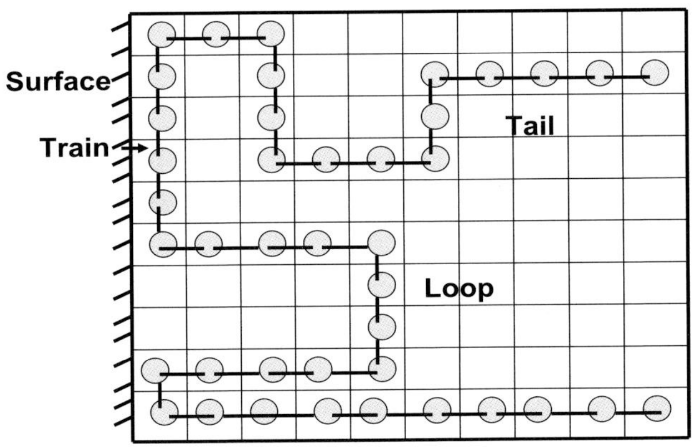

Figure 4.16 Schematic representation of a polymer molecule adsorbing on a flat surface – quasi-crystalline lattice with segments filling layers that are parallel to the surface (random mixing of segments and solvent molecules in each layer is assumed).

From the weighting factors, the statistical weight of any chain conformation can be calculated for a given segment concentration profile using a matrix formulation. Figure 4.17 shows typical adsorption isotherms plotted as surface coverage (in equivalent monolayers) versus polymer volume fraction ϕ^* in bulk solution (ϕ^* was taken to vary between 0 and 10^{-3}, which is the normal experimental range). The results in Figure 4.17 show the effect of increasing the chain

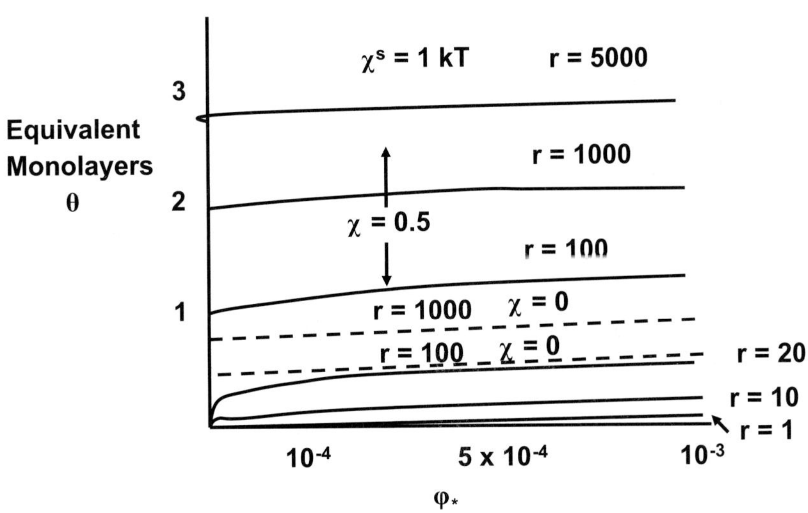

Figure 4.17 Adsorption isotherms for oligomers and polymers in the dilute region based on the SF theory.

length r and the effect of solvency (athermal solvent with $\chi = 0$ and theta solvent with $\chi = 0.5$). The adsorption energy χ^s was taken to be the same and equal to $1kT$.

When $r = 1$, θ is very small and the adsorption increases linearly with increase in ϕ^* (Henry-type isotherm). On the other hand, when $r = 10$, the isotherm deviates considerably from a straight line and approaches a Langmuirian type. However, when $r \geq 20$, high-affinity isotherms are obtained. This implies that the first added polymer chains are completely adsorbed, resulting in an extremely low polymer concentration in solution (approaching zero). This explains the irreversibility of adsorption of polymeric surfactants with $r > 100$. The adsorption isotherms with $r = 100$ and above are typical of those observed experimentally for most polymers that are not too polydisperse, i.e. showing a steep rise followed by a nearly horizontal plateau (which increases by only a few percent per decade increase in ϕ^*). In these dilute solutions, the effect of solvency is most clearly seen, with poor solvents giving the highest amounts adsorbed. In good solvents θ is much smaller and levels off for long chains to attain an adsorption plateau which is essentially independent of molecular weight.

Another point that emerges from the SF theory is the difference in shape between the experimental and theoretical adsorption isotherms in the low-concentration region. The experimental isotherms are usually rounded, whereas those predicted from theory are flat. This is accounted for in terms of the molecular weight distribution (polydispersity) which is encountered with most practical polymers. This effect has been explained by Cohen-Stuart *et al.* [28]. With polydisperse polymers, the higher molecular weight fractions adsorb preferentially over the lower ones. At low polymer concentrations, nearly all polymer fractions are adsorbed, leaving a small fraction of the polymer with the lowest molecular weights in solution. As the polymer concentration is increased, the higher molecular weight fractions displace the lower ones on the surface, which are now released in solution, thus shifting the molecular weight distribution in solution to lower values. This process continues with further increase in polymer concentration, leading to fractionation, whereby the higher molecular weight fractions are adsorbed at the expense of the lower fractions, which are released to the bulk solution. However, in very concentrated solutions, monomers adsorb preferentially with respect to polymers and short chains with respect to larger chains. This is due to the fact that in this region, the conformational entropy term predominates the free energy, disfavoring the adsorption of long chains.

According to the SF theory, the bound fraction p is high at low concentrations and relatively independent of molecular weight when $r > 20$. However, with increase in surface coverage and/or molecular weight, p tends to decrease, indicating the formation of larger loops and tails.

The structure of the adsorbed layer is described in terms of the segment density distribution. As an illustration, Figure 4.18 shows some calculations using the SF theory for loops and tails with $r = 1000$, $\phi^* = 10^{-6}$ and $\chi = 0.5$. In this example, 38% of the segments are in trains, 55.5% in loops and 6.5% in tails. This theory demonstrates the importance of tails which dominate the total distribution in the outer region.

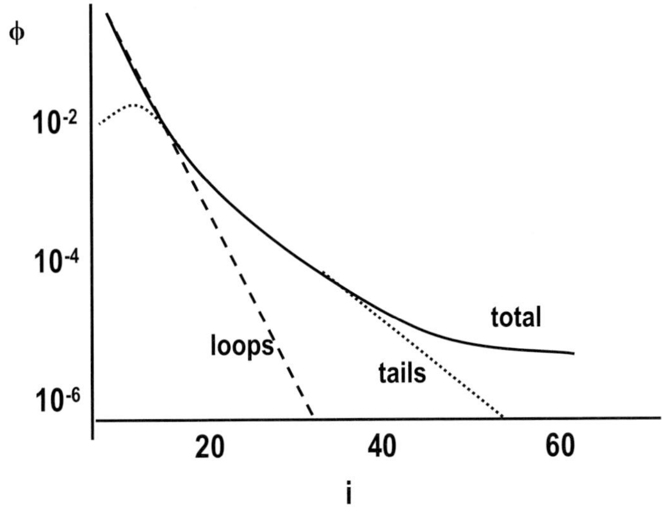

Figure 4.18 Loop, tail and total segment profile according to the SF theory.

4.3.2
Experimental Techniques for Studying Polymeric Surfactant Adsorption

As mentioned above, for full characterization of polymeric surfactant adsorption, one needs to determine three parameters: (i) the adsorbed amount Γ (mg m^{-2} or mol m^{-2}) as a function of equilibrium concentration C_{eq}, i.e. the adsorption isotherm; (ii) the fraction of segments in direct contact with the surface, p (number of segments in trains relative to the total number of segments); and (iii) the segment density distribution $\rho(z)$ or the hydrodynamic adsorbed layer thickness δ_h.

It is important to obtain the adsorption parameters as a function of the important variables of the system: (i) solvency of the medium for the chain, which can be affected by temperature and addition of salt or a non-solvent; the Flory–Huggins interaction parameter χ could be measured separately; (ii) the molecular weight of the adsorbed polymer; (iii) the affinity of the polymer to the surface as measured by the value of χ^s, the segment–surface adsorption energy; and (iv) the structure of the polymer, which is particularly important for block and graft copolymers.

4.3.2.1 Measurement of the Adsorption Isotherm

This is by far the easiest to obtain. One measures the polymeric surfactant concentration before ($C_{initial}$, C_1) and after ($C_{equilibrium}$, C_2):

$$\Gamma = \frac{(C_1 - C_2)V}{A} \tag{27}$$

where V is the total volume of the solution and A is the specific surface area (m^2 g^{-1}).

It is necessary in this case to separate the particles from the polymer solution after adsorption. This could be carried out by centrifugation and/or filtration. One should make sure that all particles are removed.

To obtain the above isotherm, one must develop a sensitive analytical technique for determination of the polymeric surfactant concentration in the ppm range. It is essential to follow the adsorption as a function of time to determine the time required to reach equilibrium. For some polymer molecules, such as poly(vinyl alcohol) (PVA) and poly(ethylene oxide) (PEO) (or blocks containing PEO), analytical methods based on complexation with iodine/potassium iodide or iodine/boric acid–potassium iodide have been established. For some polymers with specific functional groups, spectroscopic methods may be applied, e.g. UV, IR or fluorescence spectroscopy. A possible method is to measure the change in the refractive index of the polymer solution before and after adsorption. This requires very sensitive refractometers. High-resolution NMR spectroscopy has recently been applied since the polymer molecules in the adsorbed state are in a different environment to those in the bulk. The chemical shifts of functional groups within the chain are different in these two environments. This has the attraction of measuring the amount of adsorption without separating the particles.

4.3.2.2 Measurement of the Fraction of Segments p

The fraction of segments in direct contact with the surface can be directly measured using spectroscopic techniques: (i) IR if there is a specific interaction between the segments in trains and the surface, e.g. PEO on silica from non-aqueous solutions; (ii) electron spin resonance (ESR), which this requires labeling of the molecule; and (iii) NMR, pulse gradient or spin-echo NMR, which is based on the fact that the segments in trains are 'immobilized' and hence they have lower mobility than those in loops and tails.

An indirect method for the determination of p is to measure the heat of adsorption ΔH using microcalorimetry. One should then determine the heat of adsorption of a monomer H_m (or a molecule representing the monomer, e.g. ethylene glycol for PEO); p is then given by the equation

$$p = \frac{\Delta H}{H_m n} \tag{28}$$

where n is the total number of segments in the molecule.

The above indirect method is not very accurate and can only be used in a qualitative sense. It also requires very sensitive enthalpy measurements (e.g. using an LKB microcalorimeter).

4.3.2.3 Determination of the Segment Density Distribution $\rho(z)$ and Adsorbed Layer Thickness δ_h

The segment density distribution $\rho(z)$ is given by the number of segments parallel to the surface in the z-direction. The best technique to obtain $\rho(z)$ is to apply low-angle neutron scattering using the so-called contrast matching procedure. By changing the isotopic composition of the particles and medium (using ratios of deuterium to hydrogen atoms both in the particles and in the aqueous medium), one can contrast match the particle scattering length density with that of the medium.

Using the above contrast matching method, one can obtain the scattering from the polymer chain (which may be left completely protonated). In this way, one can obtain $\rho(z)$. The above technique was applied to PVA ($M = 37\,000$) on polystyrene latex and the results are shown schematically in Figure 4.19. It can be seen that $\rho(z)$ decreases exponentially with increase in z and the layer thickness extends to several tens of nanometers; this demonstrates the presence of long tails.

An alternative and more practical method to obtain the layer extension is to measure the hydrodynamic thickness of the chain, δ_h. This is defined as the distance of the plane of shear from the particle surface, and is the important parameter that determines the steric repulsion between the particles in a dispersion or emulsion droplets in an emulsion.

Several methods may be applied to obtain δ_h: (i) ultracentrifugation; (ii) dynamic light scattering (*P*hoton correlation spectroscopy, PCS); and (iii) microelectrophoresis. In all these techniques, one measures the hydrodynamic radius of the particle with and without the polymer layer and obtains δ_h by difference. For accurate determination of δ_h, one should use small particles and the layer thickness should be at least 10% of the particle radius (the accuracy of measuring the hydrodynamic radius is $\pm 1\%$). In most cases, one has to use model particles for these measurements.

The dynamic light scattering method is referred to as quasi-elastic light scattering, intensity fluctuation spectroscopy or most commonly PCS. Laser light is used to measure the intensity fluctuation of scattered light by the particles as they undergo Brownian diffusion. The sample is diluted to an extent such that

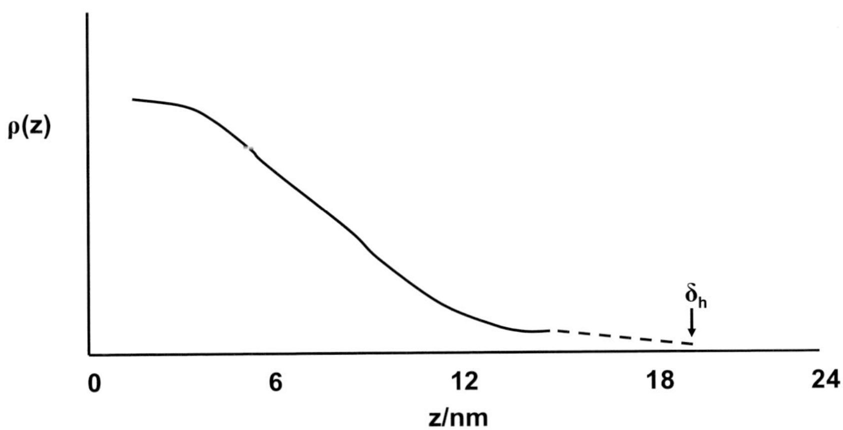

Figure 4.19 Schematic representation of the segment density distribution.

the distances between the particles become comparable to the wavelength of the laser beam (i.e. coherent over the sample) – in most modern instruments this can easily be checked. The scattering intensity of the beam is measured as a function of time to obtain the fluctuations in intensity as a result of the diffusive motion of the particles.

The above fluctuations decay exponentially with a time constant related to the diffusion coefficient D of the scatterer. From D, one can obtain the hydrodynamic radius using the Stokes–Einstein equation:

$$D = \frac{kT}{6\pi\eta R_{\mathrm{h}}} \tag{29}$$

where k is the Boltzmann constant, T is the absolute temperature and η is the viscosity of the medium.

The microelectrophoresis technique is based on measurement of the electrophoretic mobility u of the particles in the presence and absence of the polymer layer. From u, one can calculate the zeta potential ζ using the Hückel equation (which is applicable for small particles and extended double layers, i.e. $\kappa R \ll 1$, where κ is the Debye–Hückel parameter that is related to the salt concentration):

$$u = \frac{2}{3}\frac{\varepsilon\varepsilon_0\xi}{\eta} \tag{30}$$

where ε is the relative permittivity of the medium and ε_0 is the permittivity of free space.

By measuring ζ of the particles with and without the adsorbed polymer layer, one can obtain the hydrodynamic thickness δ_{h}. For accurate measurements, one should carry out the measurements at various electrolyte concentrations and extrapolate the results to the plateau value. Several automatic instruments are available for measurement of the electrophoretic mobility: the Malvern Zeta Sizer, Coulter Delsa Sizer and Broekhaven instruments. All these instruments are easy to use and a measurement can be carried out within a few minutes.

4.3.3 Examples of the Adsorption Isotherms of Nonionic Polymeric Surfactants

Figure 4.20 shows the adsorption isotherms for PEO of different molecular weights on PS (at room temperature) [29]. It can be seen that the amount adsorbed in mg m^{-2} increases with increase in the polymer molecular weight. Figure 4.21 shows the variation of the hydrodynamic thickness δ_{h} with molecular weight M. δ_{h} shows a linear increase with log M, and δ_{h} increases with n, the number of segments in the chain, according to

$$\delta_{\mathrm{h}} \approx n^{0.8} \tag{31}$$

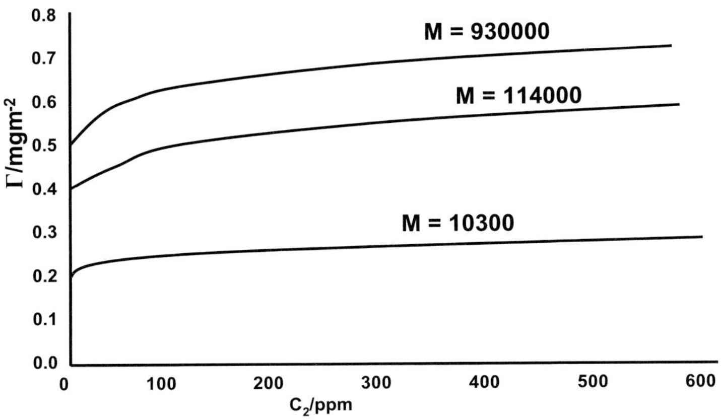

Figure 4.20 Adsorption isotherms for PEO on PS.

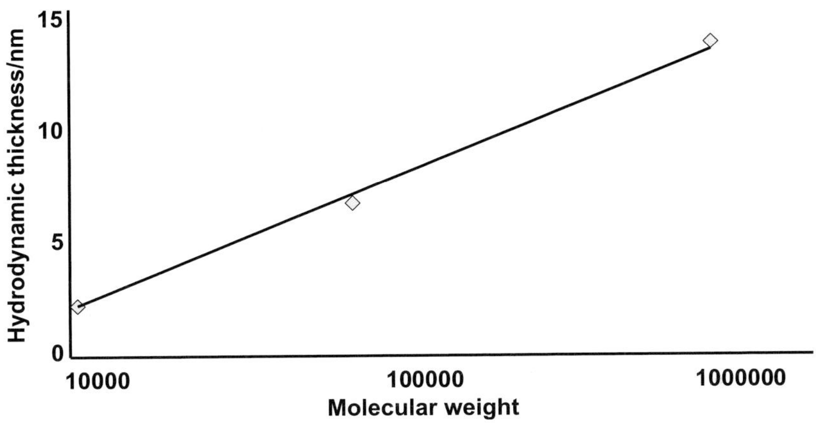

Figure 4.21 Hydrodynamic thickness of PEO on PS as a function of molecular weight.

Figure 4.22 shows the adsorption isotherms of PVA of various molecular weights on PS latex (at 25 °C) [30]. The polymers were obtained by fractionation of a commercial sample of PVA with an average molecular weight of 45 000. The polymer also contained 12% vinyl acetate groups. As with PEO, the amount of adsorption increases with increase in M. The isotherms are also of the high affinity type. Γ at the plateau increases linearly with $M^{\frac{1}{2}}$.

The hydrodynamic thickness was determined using PCS and the results are given in Table 4.6. δ_h seems to increase linearly with increase in molecular weight.

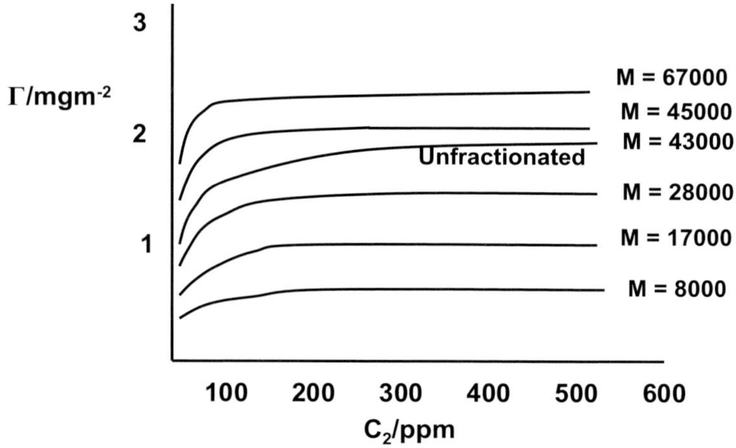

Figure 4.22 Adsorption isotherms of PVA of different molecular weights on polystyrene latex at 25 °C.

The effect of solvency on adsorption was investigated by increasing the temperature (the PVA molecules are less soluble at higher temperature) or addition of electrolyte (KCl) [31]. The results are shown in Figures 4.23 and 4.24 for $M = 65\,100$. As can be seen from Figure 4.23, an increase in temperature results in a decrease in the solvency of the medium for the chain (due to breakdown of hydrogen bonds), which results in an increase in the amount adsorbed. Addition of KCl (which reduces the solvency of the medium for the chain) results in an increase in adsorption (as predicted by theory).

The adsorption of block and graft copolymers is more complex since the intimate structure of the chain determines the extent of adsorption [29]. Random copolymers adsorb in an intermediate way with respect to the corresponding homopolymers. Block copolymers retain the adsorption preference of the individual blocks. The hydrophilic block (e.g. PEO), the buoy (previously referred to as the A chain), extends away from the particle surface into the bulk solution, whereas the

Table 4.6 Results for determination of hydrodynamic thickness.

M	δ_h/nm
67 000	25.5
43 000	19.7
28 000	14.0
17 000	9.8
8 000	3.3

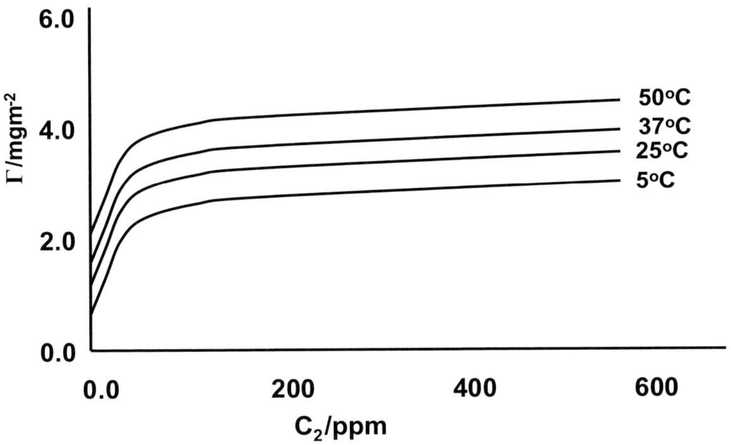

Figure 4.23 Influence of temperature on adsorption.

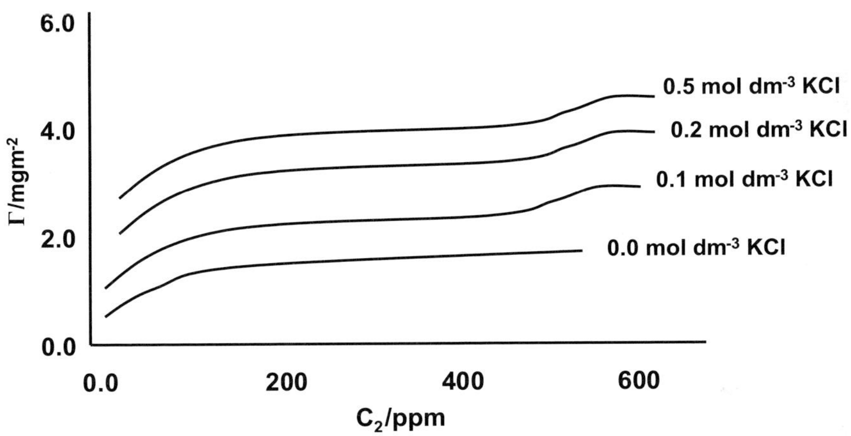

Figure 4.24 Influence of addition of KCl on adsorption.

hydrophobic anchor block (previously referred to as the B chain) (e.g. PS or PPO) provides firm attachment to the surface. Figure 4.25 shows the theoretical prediction of diblock copolymer adsorption according to the Scheutjens and Fleer theory. The surface density σ is plotted versus the fraction of anchor segments ν_A. The adsorption depends on the anchor/buoy composition.

The amount of adsorption is higher than for homopolymers and the adsorbed layer thickness is more extended and dense. Figure 4.26 shows the theoretical prediction for the adsorbed layer thickness δ, which is plotted as a function of ν_A. For a triblock copolymer A–B–A, with two buoy chains (A) and one anchor chain (B), the behavior is similar to that of diblock copolymers. This is shown in Figure 4.27 for a PEO–PPO–PEO block copolymer (Pluronic).

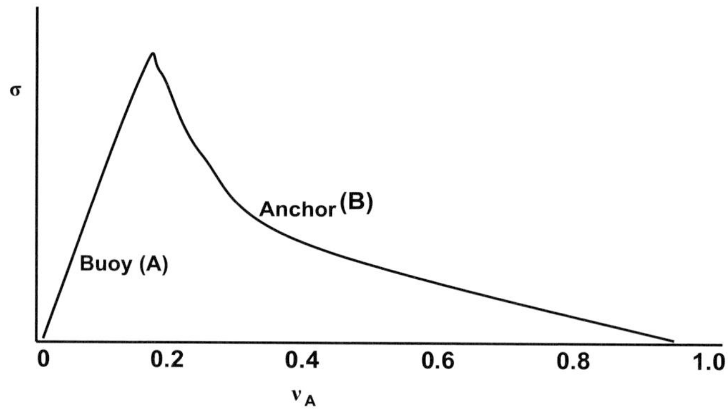

Figure 4.25 Prediction of adsorption of diblock copolymer.

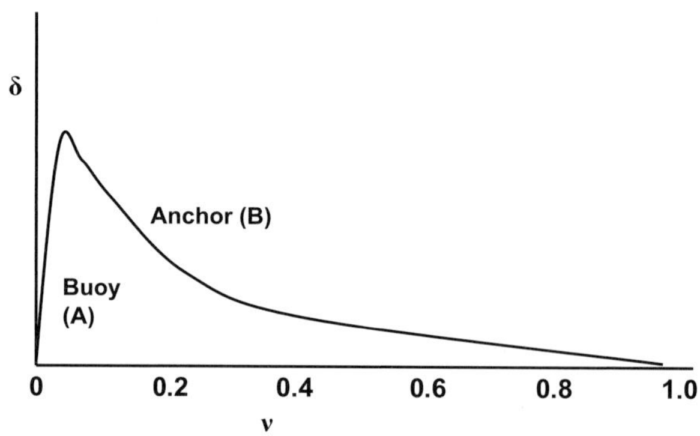

Figure 4.26 Theoretical predictions of the adsorbed layer thickness for a diblock copolymer.

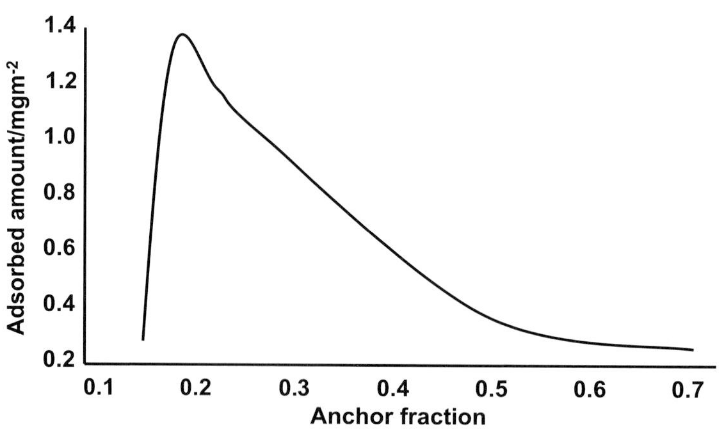

Figure 4.27 Adsorbed amount (mg m^{-2}) versus fraction of anchor segment for an A–B–A triblock copolymer (PEO–PPO–PEO).

4.4 Interaction Between Particles Containing Adsorbed Polymeric Surfactant Layers

4.4.1 Steric Stabilization

When two particles, each with a radius R and containing an adsorbed polymer layer with a hydrodynamic thickness δ_h, approach each other to a surface–surface separation distance h that is smaller than $2\delta_h$, the polymer layers interact with each other, resulting in two main situations [32]: (i) the polymer chains may overlap with each other; or (ii) the polymer layer may undergo some compression. In both cases, there will be an increase in the local segment density of the polymer chains in the interaction region. This is illustrated schematically in Figure 4.28. The real situation is perhaps in between the two cases, i.e. the polymer chains may undergo some interpenetration and some compression.

Provided that the dangling chains (the A chains in A–B, A–B–A block or BA_n graft copolymers) are in a good solvent, this local increase in segment density in the interaction zone will result in strong repulsion as a result of two main effects: (i) an increase in the osmotic pressure in the overlap region as a result of the unfavorable mixing of the polymer chains, when these are in good solvent conditions; this is referred to as osmotic repulsion or mixing interaction and is described by a free energy of interaction G_{mix}; and (ii) a reduction in the configurational entropy of the chains in the interaction zone; this entropy reduction results from the decrease in the volume available for the chains when these are either overlapped or compressed; this is referred to as volume restriction, entropic or elastic interaction and is described by a free energy of interaction G_{el}.

Combination of G_{mix} and G_{el} is usually referred to as the steric interaction free energy, G_s, i.e.

$$G_s = G_{mix} + G_{el} \tag{32}$$

The sign of G_{mix} depends on the solvency of the medium for the chains. If in a good solvent, i.e. the Flory–Huggins interaction parameter $\chi < 0.5$, then G_{mix} is

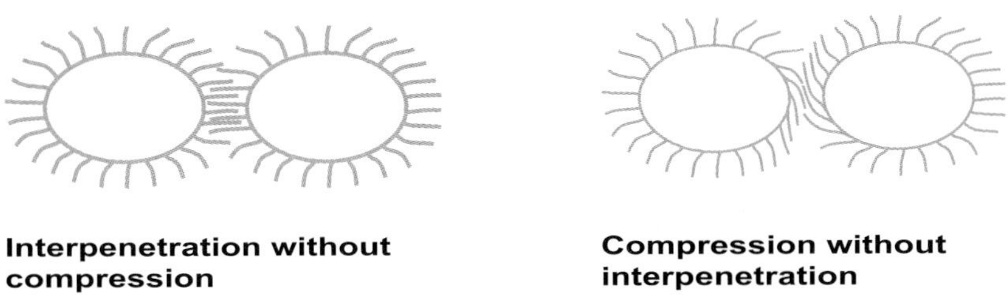

Figure 4.28 Schematic representation of the interaction between particles containing adsorbed polymer layers.

positive and the mixing interaction leads to repulsion (see below). In contrast, if $\chi > 0.5$ (i.e. the chains are in a poor solvent condition), G_{mix} is negative and the mixing interaction becomes attractive. G_{el} is always positive and hence in some cases one can produce stable dispersions in a relatively poor solvent (enhanced steric stabilization).

4.4.2
Mixing Interaction G_{mix}

This results from the unfavorable mixing of the polymer chains, when these are in good solvent conditions. This is shown schematically in Figure 4.29.

Consider two spherical particles with the same radius and each containing an adsorbed polymer layer with thickness δ. Before overlap, one can define in each polymer layer a chemical potential for the solvent μ_i^{α} and a volume fraction for the polymer in the layer ϕ_2. In the overlap region (volume element dV), the chemical potential of the solvent is reduced to μ_i^{β}. This results from the increase in polymer segment concentration in this overlap region.

In the overlap region, the chemical potential of the polymer chains is now higher than in the rest of the layer (with no overlap). This amounts to an increase in the osmotic pressure in the overlap region and, as a result, solvent will diffuse from the bulk to the overlap region, thus separating the particles, and hence a strong repulsive energy arises from this effect. The above repulsive energy can be calculated by considering the free energy of mixing of two polymer solutions, as for example treated by Flory and Krigbaum [33]. The free energy of mixing is given by two terms: (i) an entropy term that depends on the volume fraction of polymer and solvent and (ii) an energy term that is determined by the Flory–Huggins interaction parameter χ.

Using the above theory, one can derive an expression for the free energy of mixing of two polymer layers (assuming a uniform segment density distribution in each layer) surrounding two spherical particles as a function of the separation distance h between the particles. The expression for G_{mix} is [34]

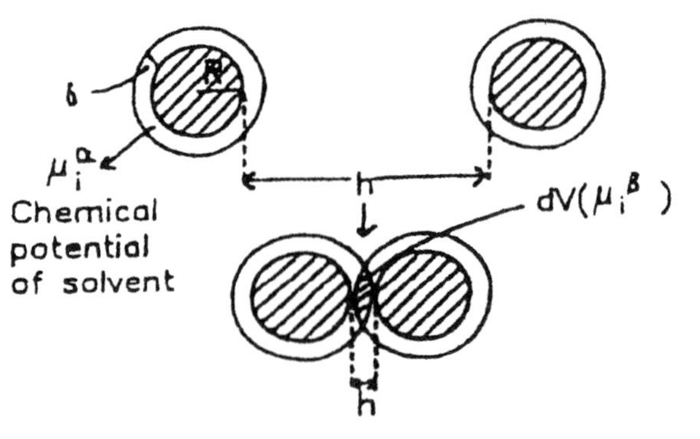

Figure 4.29 Schematic representation of polymer layer overlap.

$$\frac{G_{\mathrm{mix}}}{kT} = \left(\frac{2V_2{}^2}{V_1}\right)\nu_2{}^2\left(\frac{1}{2}-\chi\right)\left(\delta-\frac{h}{2}\right)^2\left(3R+2\delta+\frac{h}{2}\right) \tag{33}$$

where k is the Boltzmann constant, T is the absolute temperature, V_2 is the molar volume of polymer, V_1 is the molar volume of solvent and ν_2 is the number of polymer chains per unit area.

The sign of G_{mix} depends on the value of the Flory–Huggins interaction parameter χ: if $\chi < 0.5$, G_{mix} is positive and the interaction is repulsive; if $\chi > 0.5$, G_{mix} is negative and the interaction is attractive; if $\chi = 0.5$, $G_{\mathrm{mix}} = 0$ and this defines the θ-condition.

4.4.3 Elastic Interaction G_{el}

This arises from the decrease in configurational entropy of the chains on the approach of a second particle. As a result of this approach, the volume available for the chains becomes restricted, resulting in a decrease in the number of configurations. This can be illustrated by considering a simple molecule, represented by a rod that rotates freely in a hemisphere across a surface (Figure 4.30). When the two surfaces are separated by an infinite distance ∞, the number of configurations of the rod is $\Omega(\infty)$, which is proportional to the volume of the hemisphere. When a second particle approaches to a distance h such that it cuts the hemisphere (losing some volume), the volume available to the chains is reduced and the number of configurations become $\Omega(h)$ which is less than $\Omega(\infty)$. For two flat plates, G_{el} is given by the following expression [35]:

$$\frac{G_{\mathrm{el}}}{kT} = 2\nu_2 \ln\left[\frac{\Omega(h)}{\Omega(\infty)}\right] = 2\nu_2 R_{\mathrm{el}}(h) \tag{34}$$

where $R_{\mathrm{el}}(h)$ is a geometric function whose form depends on the segment density distribution. It should be stressed that G_{el} is always positive and could play a major role in steric stabilization. It becomes very strong when the separation distance between the particles becomes comparable to the adsorbed layer thickness δ.

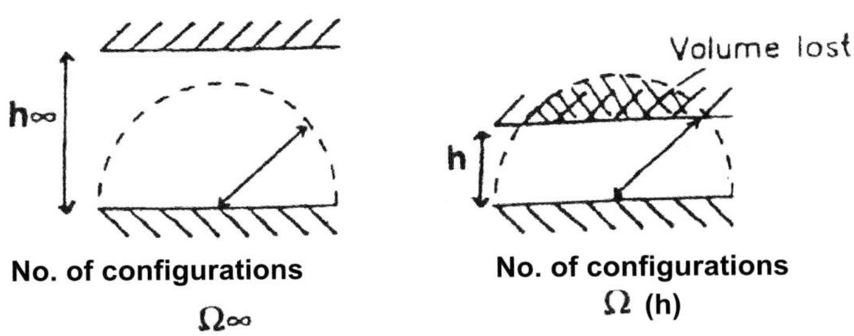

Figure 4.30 Schematic representation of configurational entropy loss on approach of a second particle.

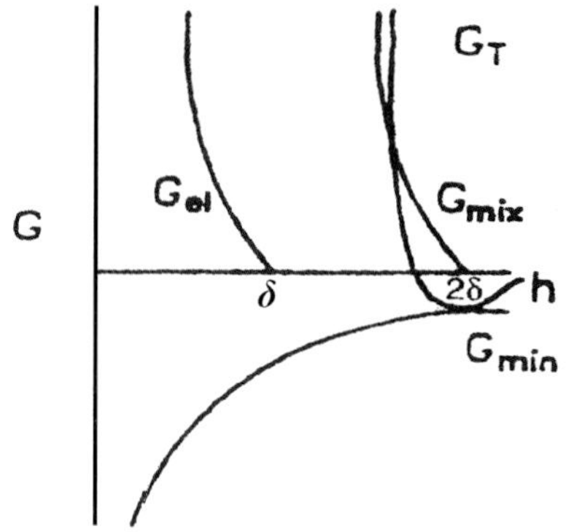

Figure 4.31 Energy–distance curves for sterically stabilized systems.

Combination of G_{mix} and G_{el} with G_A (the van der Waals attractive energy) gives the total free energy of interaction G_T (assuming there is no contribution from any residual electrostatic interaction) [36], i.e.

$$G_T = G_{mix} + G_{el} + G_A \tag{35}$$

A schematic representation of the variation of G_{mix}, G_{el}, G_A and G_T with surface–surface separation distance h is shown in Figure 4.31.

G_{mix} increases very sharply with decrease in h when $h < 2\delta$. G_{el} increases very sharply with decrease in h when $h < \delta$. G_T versus h shows a minimum, G_{min}, at separation distances comparable to 2δ. When $h < 2\delta$, G_T shows a rapid increase with decrease in h.

The depth of the minimum depends on the Hamaker constant A, the particle radius R and adsorbed layer thickness δ. G_{min} increases with increase in A and R.

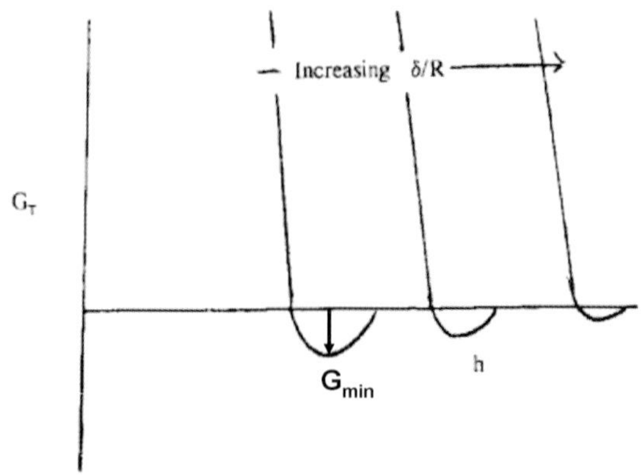

Figure 4.32 Variation of G_{min} with δ/R.

At a given A and R, G_{min} increases with decrease in δ (i.e. with decrease in the molecular weight, M, of the stabilizer). This is illustrated in Figure 4.32, which shows the energy–distance curves as a function of δ/R. The larger the value of δ/R, the smaller is the value of G_{min}. In this case, the system may approach thermodynamic stability as is the case with nano-dispersions.

4.4.4 Criteria for Effective Steric Stabilization

The criteria are as follows:

1. The particles should be completely covered by the polymer (the amount of polymer should correspond to the plateau value). Any bare patches may cause flocculation either by van der Waals attraction (between the bare patches) or by bridging flocculation (whereby a polymer molecule will become simultaneously adsorbed on two or more particles).
2. The polymer should be strongly 'anchored' to the particle surfaces, to prevent any displacement during particle approach. This is particularly important for concentrated suspensions. For this purpose, A–B, A–B–A block and BA_n graft copolymers are the most suitable where the chain B is chosen to be highly insoluble in the medium and to have a strong affinity to the surface. Examples of B groups for hydrophobic particles in aqueous media are polystyrene and poly(methyl methacrylate).
3. The stabilizing chain A should be highly soluble in the medium and strongly solvated by its molecules. Examples of A chains in aqueous media are poly(ethylene oxide) and poly(vinyl alcohol).
4. δ should be sufficiently large (>10 nm) to prevent weak flocculation.

4.4.5 Flocculation of Sterically Stabilized Dispersions

Three main types of flocculation may be distinguished, as follows.

Weak Flocculation This occurs when the thickness of the adsorbed layer is small (usually <5 nm), particularly when the particle radius and Hamaker constant are large.

Depletion Flocculation This is obtained by addition of 'free' (non-adsorbing) polymer in the continuous phase [37]. At a critical concentration, or volume fraction of free polymer, φ_p^+, weak flocculation occurs, since the free polymer coils become 'squeezed out' from between the droplets. This is illustrated in Figure 4.33, which shows the situation when the polymer volume fraction exceeds the critical concentration.

The osmotic pressure outside the particles is higher than in between the particles, which results in an attraction whose magnitude depends on the concentration

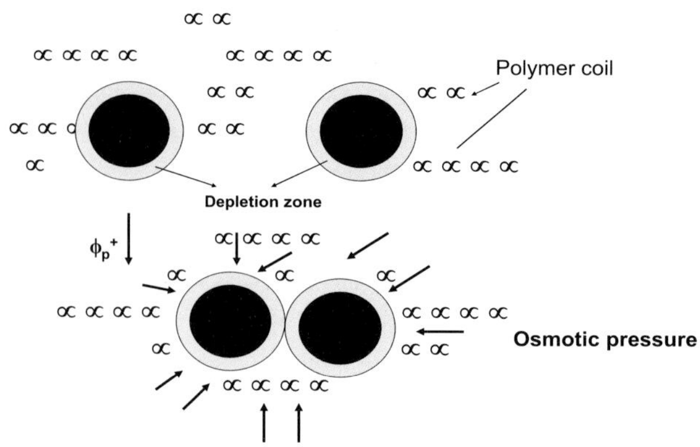

Figure 4.33 Schematic representation of depletion flocculation.

of the free polymer and its molecular weight, in addition to the droplet size and φ. The value of $\varphi_p{}^+$ decreases with increase in the molecular weight of the free polymer. It also decreases as the volume fraction of the dispersion increases.

The above weak flocculation can be applied to reduce sedimentation, although it suffers from the drawbacks of temperature dependence: as the temperature increases, the hydrodynamic radius of the free polymer decreases (due to dehydration) and hence more polymer will be required to achieve the same effect at lower temperatures. Further, if the free polymer concentration is increased above a certain limit, phase separation may occur and the flocculated particles may sediment faster than in the absence of the free polymer.

Incipient Flocculation This occurs when the solvency of the medium is reduced to become worse than θ-solvent (i.e. $\chi > 0.5$). This is illustrated in Figure 4.34, where χ was increased from <0.5 (good solvent) to >0.5 (poor solvent).

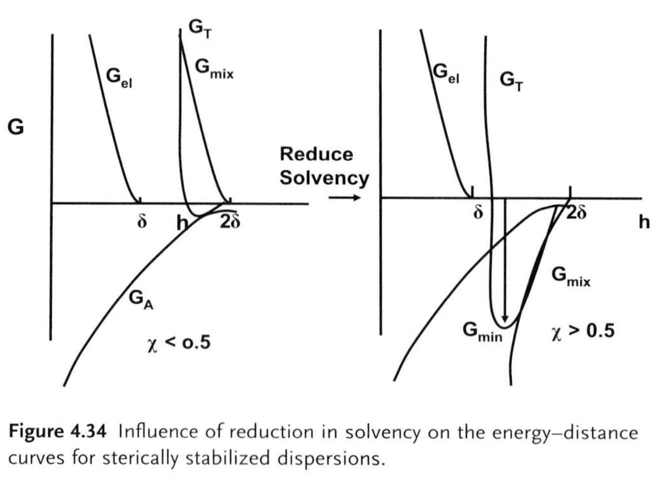

Figure 4.34 Influence of reduction in solvency on the energy–distance curves for sterically stabilized dispersions.

When $\chi > 0.5$, G_{mix} becomes negative (attractive), which when combined with the van der Waals attraction at this separation distance gives a deep minimum, causing flocculation. In most cases, there is a correlation between the critical flocculation point and the θ-condition of the medium. A good correlation is found in many cases between the critical flocculation temperature (CFT) and θ-temperature of the polymer in solution (with block and graft copolymers one should consider the θ-temperature of the stabilizing chains A) [32]. A good correlation is also found between the critical volume fraction (CFV) of a non-solvent for the polymer chains and their θ-point under these conditions. However, in some cases such a correlation may break down, particularly the case for polymers which adsorb by multi-point attachment. This situation was described by Napper [32], who referred to it as 'enhanced' steric stabilization.

Thus, by measuring the θ-point (CFT or CFV) for the polymer chains (A) in the medium under investigation (which could be obtained from viscosity measurements), one can establish the stability conditions for a dispersion, before its preparation. This procedure also helps in designing effective steric stabilizers such as block and graft copolymers.

References

1 J. Lyklema, Structure of the solid/liquid interface and the electrical double layer, in *Solid/Liquid Dispersions*, ed. Th.F. Tadros, Academic Press, London (1987).
2 H. van Olphen, *Clay Colloid Chemistry*, John Wiley & Sons, Inc., New York (1963).
3 G. Gouy, *J. Phys.*, 9 (4), 457 (1910); *Ann. Phys.*, *7* (9), 129 (1917).
4 D.L. Chapman, *Philos. Mag.*, *25* (6), 475 (1913).
5 O. Stern, *Z. Elektrochem.*, *30*, 508 (1924).
6 B.H. Bijesterbosch, Stability of solid/liquid dispersions, in *Solid/Liquid Dispersions*, ed. Th.F. Tadros, Academic Press, London (1987).
7 H.R. Kruyt (ed.), *Colloid Science*, Vol. I, Elsevier, Amsterdam (1952).
8 J. Lyklema, *Fundamentals of Interface and Colloid Science*, Vol. 1, Academic Press, London (1991).
9 B.V. Deryaguin and L. Landau, *Acta Physicochim. URSS*, *14*, 633 (1941).
10 E.J.W. Verwey and J.Th.G. Overbeek, *Theory of Stability of Lyophobic Colloids*, Elsevier, Amsterdam (1948).
11 M. von Smoluchowski, *Phys. Z.*, *17*, 557, 585 (1917); *Z. Phys. Chem.*, *92*, 129 (1927).
12 N. Fuchs, *Z. Phys.*, *89*, 736 (1936).
13 H. Reerink and J.Th.G. Overbeek, *Discuss. Faraday Soc.*, *18*, 74 (1954).
14 R.J. Hunter, *Zeta Potential in Colloid Science; Principles and Applications*, Academic Press, London (1981).
15 M. von Smoluchowski, *Handbuch der Electricität und des Magnetismus*, Vol. II, Barth, Leipzig (1914).
16 E. Hückel, *Phys. Z.*, *25*, 204 (1924).
17 D.C. Henry, *Proc. R. Soc. London, Ser. A*, *133*, 106 (1948).
18 P.H. Wiersema, A.L. Loeb and J.Th.G. Overbeek, *J. Colloid Interface Sci.*, *22*, 78 (1967).
19 R.H. Ottewill and J.N. Shaw, *J. Electroanal. Interfacial Electrochem.*, *37*, 133 (1972).
20 Th.F. Tadros, Polymer adsorption and dispersion stability, in *The Effect of Polymers on Dispersion Properties*, ed. Th.F. Tadros, Academic Press, London (1981).
21 A. Silberberg, *J. Chem. Phys.*, *48*, 2835 (1968).
22 C.A. Hoeve, *J. Polym. Sci.*, *30*, 361 (1970); *34*, 1 (1971).
23 R.J. Roe, *J. Chem. Phys.*, *60*, 4192 (1974).
24 J.M.H.M. Scheutjens and G.J. Fleer, *J. Phys. Chem.*, *83*, 1919 (1979).

25 J.M.H.M. Scheutjens and G.J. Fleer, *J. Phys. Chem.*, *84*, 178 (1980).

26 J.M.H.M. Scheutjens and G.J. Fleer, *Adv. Colloid Interface Sci.*, *16*, 341 (1982).

27 G.J. Fleer, M.A. Cohen-Stuart, J.M.H.M. Scheutjens, T. Cosgrove and B. Vincent, *Polymers at Interfaces*, Chapman and Hall, London (1993).

28 M.A. Cohen-Stuart, G.J. Fleer and B.H. Bijesterbosch, *J. Colloid Interface Sci.*, *90*, 321 (1982).

29 T.M. Obey and P.C. Griffiths, in *Principles of Polymer Science and Technology in Cosmetics and Personal Care*, ed. E.D. Goddard and J.V Gruber, Marcel Dekker, New York, Chapter 2 (1999).

30 M.J. Garvey, Th.F. Tadros and B. Vincent, *J. Colloid Interface Sci.*, *49*, 57 (1974).

31 Th. van den Boomgaard, T.A. King, Th.F. Tadros, H. Tang and B. Vincent, *J. Colloid Interface Sci.*, *61*, 68 (1978).

32 D.H. Napper, *Polymeric Stabilisation of Colloidal Dispersions*, Academic Press, London (1983).

33 P.J. Flory and W.R. Krigbaum, *J. Chem. Phys.*, *18*, 1086 (1950).

34 E.W. Fischer, *Kolloid Z.*, *160*, 120 (1958).

35 E.L. Mackor and J.H. van der Waals, *J. Colloid Sci.*, *7*, 535 (1951).

36 F.Th. Hesselink, A. Vrij and J.Th.G. Overbeek, *J. Phys. Chem.*, *75*, 2094 (1971).

37 S. Asakura and F. Oosawa, *J. Polym. Sci.*, *33*, 245 (1958).

5
Particle Deposition and Adhesion

5.1
Deposition of Particles on Surfaces

The deposition of particles on surfaces is an important process in paint application, since it governs the paint film properties. Two main types of particles must be considered, namely the latex particles (which may be considered as emulsions) that determine film formation and the pigment particles such as TiO_2 and other colored pigments. Particle deposition is followed by their adhesion and it is essential to have uniform deposition (without any flocculation) to ensure that the final adhered film has the required properties such as gloss and durability. Particle deposition is determined by long-range forces: van der Waals attraction, electrostatic repulsion or attraction and the presence of adsorbed or grafted surfactants, polymers or polyelectrolytes (referred to as steric interaction). The first part of this chapter discusses the role of van der Waals attraction and electrostatic repulsion (or attraction) on particle deposition.

5.1.1
Van der Waals Attraction

This has been described in detail in Chapter 4, but for the sake of completion of this section a summary will be given. The attraction between atoms or molecules is of three types: dipole–dipole (Keesom force), dipole–induced dipole (Debye force) and London dispersion forces.

The most important is the London dispersion attraction, which operates for polar and non-polar atoms and molecules. This attractive energy is of short range and is inversely proportional to the sixth power of the distance between the atoms or molecules.

For an assembly of atoms or molecules, the London attraction may be summed, resulting in significant attraction that operates over a large distance of separation h between the particles. For two identical particles with radius a, the attractive energy G_A is given by the simple expression [1]

$$G_A = -\frac{Aa}{12h} \tag{1}$$

Colloids and Interface Science Series, Vol. 6
Colloids in Paints. Tharwat F. Tadros

ISBN: 978-3-527-31466-9

where A is the effective Hamaker constant, which is given by

$$A = (A_{11}^{\frac{1}{2}} - A_{22}^{\frac{1}{2}})^2 \tag{2}$$

where A_{11} is the Hamaker constant of the particles and A_{22} is the effective Hamaker constant of the medium.

The Hamaker constant of any substance is given by the expression

$$A = \pi q^2 \beta \tag{3}$$

where q is the number of atoms or molecules per unit volume and β is the London dispersion constant.

For two different particles or particle and surface with Hamaker constants A_{11} and A_{22} separated by a medium with Hamaker constant A_{33}, the effective Hamaker constant A is given by

$$A = (A_{11}^{\frac{1}{2}} - A_{33}^{\frac{1}{2}})(A_{22}^{\frac{1}{2}} - A_{33}^{\frac{1}{2}}) \tag{4}$$

A schematic representation of the variation of G_A with h is given in Figure 5.1. It can be seen that G_A increases with decrease in h and it reaches very high values at very small h values. At extremely short h, G_A increases (Born repulsion).

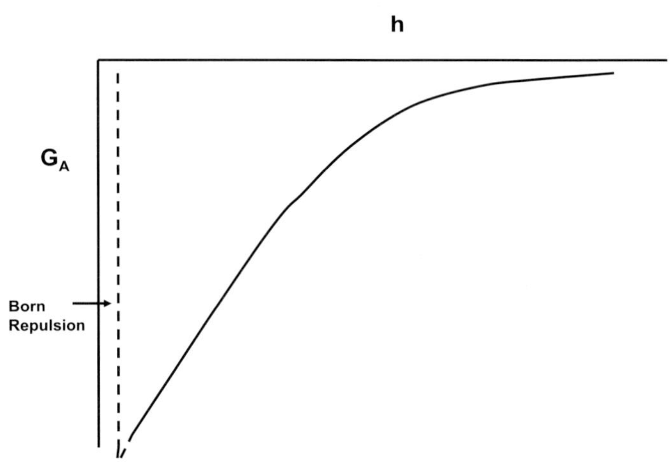

Figure 5.1 Schematic representation of the variation of G_A with h.

5.1.2
Electrostatic Repulsion

Electrostatic repulsion occurs as a result of the presence of an electrical double layer consisting of a surface charge that is compensated for by an unequal distri-

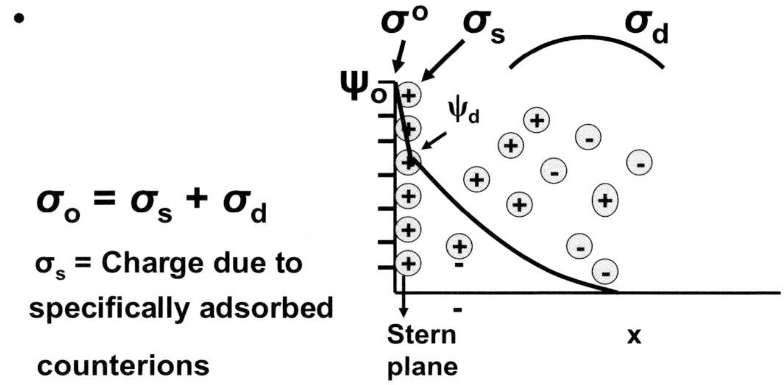

Figure 5.2 Schematic representation of the double layer.

bution of counter- and co-ions. A schematic representation of the double layer is shown in Figure 5.2 for a negatively charged surface [2, 3].

The surface potential ψ_0 decreases linearly to ψ_δ, the Stern potential, which is nearly equal to the measurable zeta (ζ) potential.

The double layer extension depends on the electrolyte concentration and valency of the ions as given by $1/\kappa$, the 'thickness of the double layer'. The lower the electrolyte concentration and the lower the valency of the ions, the larger is the value of $1/\kappa$. For example, for a 1:1 electrolyte (e.g. NaCl), $1/\kappa = 100$ nm at 10^{-5} mol dm^{-3}, 10 nm at 10^{-3} mol dm^{-3} and 1 nm at 10^{-1} mol dm^{-3}.

When two particles of the same double layer sign approach to a distance of separation h that is less than twice the double layer thickness, repulsion occurs, since the double layers begin to overlap. The repulsion between particles or between a particle and a surface decreases with increase in electrolyte concentration. This is shown schematically in Figure 5.3, where G_{el} is plotted versus h at low and high electrolyte concentration.

Combination of G_A and G_{el} at various h values results in the energy–distance curve illustrated in Figure 5.4, which forms the basis of the theory of colloid stability due to Deryaguin–Landau–Verwey–Overbeek (DLVO theory) [4, 5]. The energy–distance curve is characterized by two minima, a shallow secondary minimum (weak and reversible attraction) and a primary deep minimum (strong and irreversible attraction).

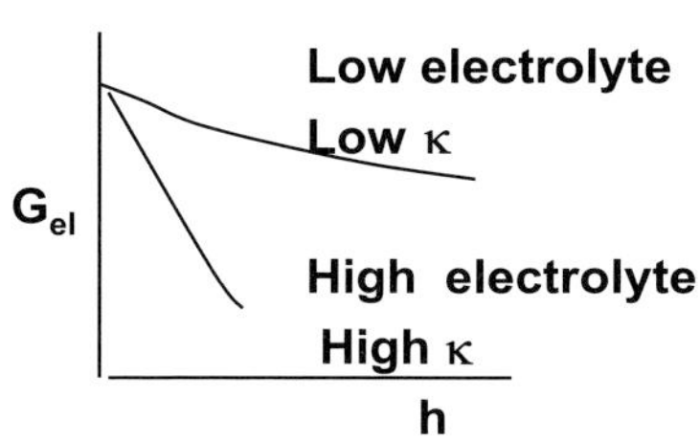

Figure 5.3 Variation of G_{el} with h at low and high electrolyte concentration.

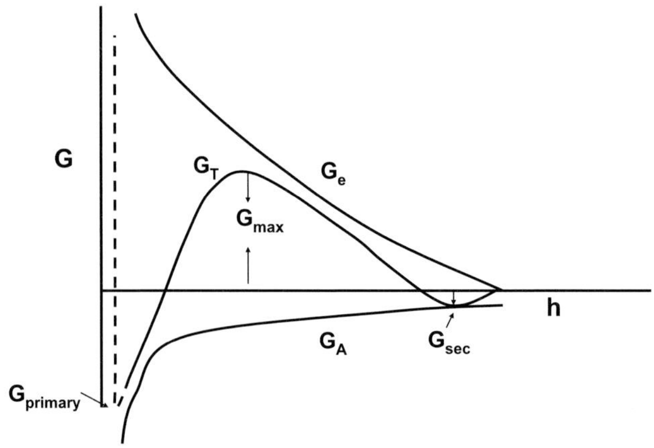

Figure 5.4 Energy–distance curve (DLVO theory).

Particles deposited under conditions of secondary minimum will be weakly attached, whereas particles deposited under conditions of primary minimum will be strongly attached. At intermediate distances of separation, an energy maximum is obtained whose height depends on the surface or zeta potential, electrolyte concentration and valency of the ions. This maximum prevents particle deposition.

The magnitudes of the energy minima and the energy maximum depend on electrolyte concentration and valency, as illustrated in Figure 5.5 for a 1:1 electro-

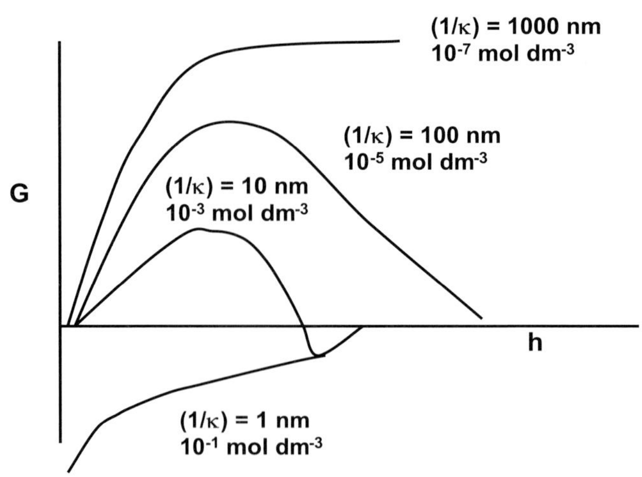

Figure 5.5 G_T versus h curves at various NaCl concentrations.

lyte (e.g. NaCl) at various concentrations. It can be seen that G_{max} decreases with increase in NaCl concentration and eventually it disappears at 10^{-1} mol dm^{-3}. Thus, particle deposition for particles with the same sign as the surface will increase with increase in electrolyte concentrations.

The above trend was confirmed by Hull and Kitchener [6] using a rotating disc coated with a negative film and negative polystyrene latex particles. The number of polystyrene particles deposited was found to increase with increase in NaCl concentration, reaching a maximum at $C_{NaCl} > 10^{-1}$ mol dm^{-3}. The ratio of the maximum number of particles deposited N_{max} to the number deposited at any other NaCl concentration N_d (the so-called stability ratio W) was calculated and plotted versus NaCl concentration:

$$W = \frac{N_{max}}{N_d} \tag{5}$$

Figure 5.6 illustrates such plots and clearly shows that W decreases with increase in NaCl concentration, reaching a minimum above 10^{-3} mol dm^{-3}, whereby maximum deposition occurs. Similar results were obtained by Tadros [7] using a rotating cylinder apparatus and the results are shown in Figure 5.7.

It can be concluded from the above results that the deposition of particles on substrates of the same sign will increase with increase in electrolyte concentration. However, the situation with an oppositely charged surface to the particles being deposited is very different. In this case, attraction between oppositely charged surfaces will occur, a phenomenon referred to as heteroflocculation. This is illustrated schematically in Figure 5.8 for positively charged polystyrene latex particles on a negative surface; both surfaces were covered with a nonionic polymer layer.

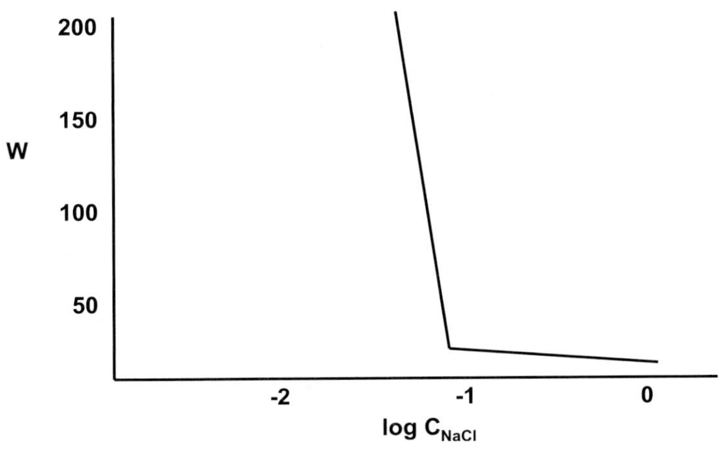

Figure 5.6 Variation of W with log C_{NaCl} using rotating disc.

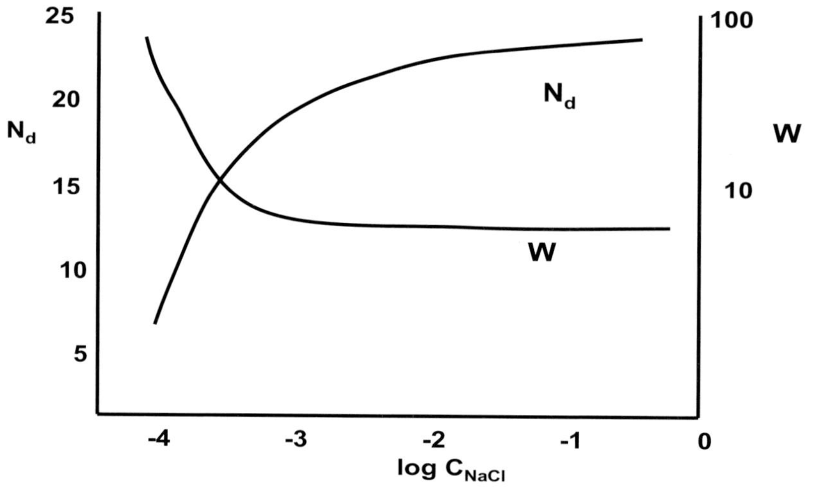

Figure 5.7 N_d and W versus C_{NaCl} using a rotating cylinder.

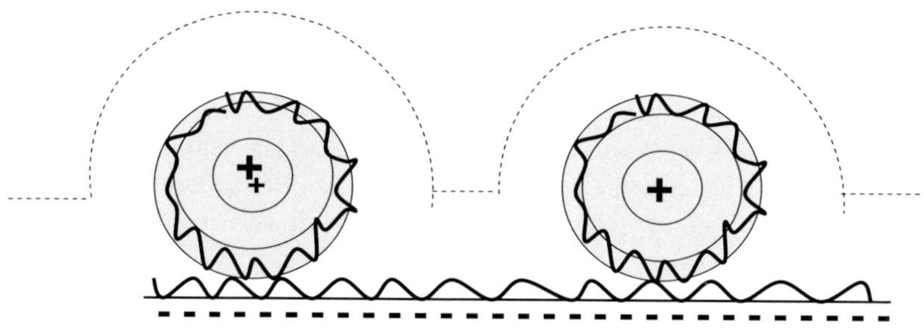

Figure 5.8 Deposition of positively charged particles on negatively charged surface.

The effect of addition of electrolyte in this case will be the opposite of that observed with surfaces of the same charge. Attraction between oppositely charged double layers will be higher at lower electrolyte concentrations. In other words, addition of electrolyte in this case will decrease deposition.

5.1.3
Effect of Polymers and Polyelectrolytes on Particle Deposition

Polymers and polyelectrolytes, of both the natural and synthetic types, are commonly used in most paint formulations. These materials are used as thickening agents, film formers, resinous powder and humectants. For example, thickening agents, sometimes referred to as rheology modifiers, are used in many paint formulations to maintain the product stability.

In many formulations, polymers and surfactants are present and interaction between them can produce remarkable effects. Several structures can be identified and the aggregates produced can have profound effects on particle deposition. With many paint formulations, polymers are added and these are mostly polyelectrolytes with cationic charges, which are essential for strong attachment to the negatively charged surface.

For convenience, the effects of three classes of polymers on particle deposition will be considered separately: nonionic polymers, anionic polyelectrolytes and cationic polyelectrolytes.

5.1.4
Effect of Nonionic Polymers on Particle Deposition

Nonionic polymers can be of the synthetic type such as polyvinylpyrrolidone or natural such as many polysaccharides. The role of nonionic polymers in particle deposition depends on the manner in which they interact with the surface and particle to be deposited. With many high molecular weight polymers, the chains adopt a conformation forming loops and tails that may extend several nanometers from the surface. If there is not sufficient polymer to cover the surfaces fully, bridging may occur, resulting in enhancement of deposition. In contrast, if there is sufficient polymer to cover both surfaces, the loops and tails provide steric repulsion, resulting in a reduction in deposition.

One may be able to correlate particle deposition with the adsorption isotherm. Under conditions of incomplete coverage, i.e. well before the plateau value is reached, particle deposition is enhanced. Under conditions of complete coverage,

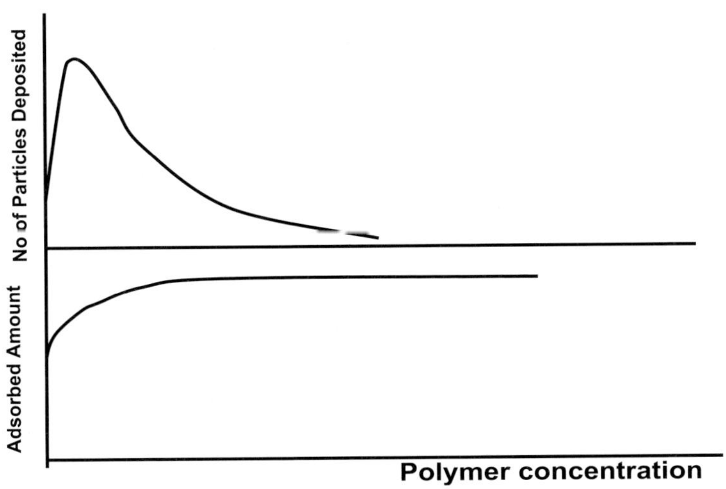

Figure 5.9 Correlation of particle deposition with polymer adsorption.

one observes a reduction in deposition, and at sufficient coverage deposition may be prevented altogether. This is illustrated schematically in Figure 5.9, which shows the correlation of the adsorption isotherm with particle surface deposition.

The most commonly used nonionic polymers in paints are polysaccharide based. Polysaccharides perform a number of functions in paints: rheology modifiers, suspending agents and emulsifying agents. Polysaccharides are sometimes referred to as 'polyglycans' or 'hydrocolloids'. The majority of polysaccharides are comprised primarily of six-membered cyclic structures known as pyranose rings (five carbon atoms and one oxygen atom). Many polysaccharides form helices, which are a tertiary spatial configuration, arranged to minimize the total energy of the polysaccharide (e.g. xanthan gum).

The behavior of polysaccharides is critically influenced by the nature of the substituent groups bound to the individual monosaccharides (natural or synthetic). Anionic charges may also occur in natural polysaccharides and this will have a large influence on the adsorption and conformation of the polymer chain.

The effect of polysaccharides on particle deposition is rather complex and depends on the structure of the molecule and interaction with other ingredients in the formulation.

5.1.5
Effect of Anionic Polymers on Particle Deposition

Many paint formulations contain anionic polymers, mostly of the polyacrylate and polysaccharide type. The role of the anionic polymers is particle deposition is complex since these polyelectrolytes interact with ions in the formulation, e.g. Ca^{2+}, and also with the surfactants used. Two of the most commonly used anionic polysaccharides are carboxymethylcellulose and carboxymethyllecithin, obtained by carboxymethylation of cellulose and chitin, respectively.

Several naturally occurring anionic polysaccharides exist, e.g. alginic acid, pectin, carragenans, xanthan gum, hyaluronic acid and gum exudates (arabic gum, karaya, tragacanth, etc.). Cross-linking sites that occur when a polyvalent cation (e.g. Ca^{2+}) causes interpolysaccharide binding are called 'junction zones'.

The above complexes which may produce colloidal particles will have a large influence on the deposition of other particles in the formulation. They may enhance binding, simply by a cooperative effect in which the polysaccharide complex interacts with the particles and increases the attraction to the surface. The pH of the whole system plays a major role since it affects the dissociation of the carboxylic groups.

Many of the anionic polysaccharides and their complexes affect the rheology of the system and this has a pronounced effect on particle deposition. Any increase in the viscosity of the system will reduce the flux of the particles to the surface, which may reduce particle deposition. This reduction may be offset by specific interactions between the particles and the polyanion or its complex.

5.1.6 Effect of Cationic Polymers on Particle Deposition

Polycationic polyelectrolytes have a pronounced effect on particle deposition due to their interaction with the substrate and the particles. One of the earliest polycationic polymers was polyethylenimine (PEI), which was used in some paint formulations. Later, an important class of cellulosic polycationic polymers was introduced with the trade name Polymer JR (Amerchol Corporation). Other synthetic polycationic polymers from Calgon Corporation are Merquat 100 (based on dimethyldiallylamine chloride) and Merquat 550 (based acrylamide–dimethyldiallylamine chloride).

Several naturally occurring polycationic polymers exist, such as chitosan (polyglycan with cationic charges) that is positively charged at pH < 7, cationic hydroxyethylcellulose and cationic guar gum. These polycationic polymers interact with anionic surfactants present in the formulation and at a specific surfactant concentration a rapid increase in the viscosity of the solution is observed. At higher surfactant concentrations, precipitation of the polymer–surfactant complex occurs, and at even higher surfactant concentrations, repeptization may occur.

Clearly, the above interactions will have pronounced effects on particle deposition. In the absence of any other effects, addition of cationic polyelectrolytes can enhance particle deposition either by simple charge neutralization or by 'bridging' between the particle and the surface. At high polyelectrolyte concentrations, when there are sufficient molecules to coat both particle and surface, repulsion may occur, resulting in a decrease in deposition.

However, the above effects are complicated by the interaction of the polycationic polymer with surfactants in the formulation, which complicates predictions regarding particle deposition. Investigations of the interactions that take place between the polycation, surfactants and other ingredients in the formulation are essential before a complete picture of particle deposition is possible.

5.2 Particle–Surface Adhesion

Adhesion is the force necessary to separate adherents; it is governed by short-range forces [8, 9]. Adhesion is more complex than deposition and more difficult to measure. No quantitative theory is available that can describe all adhesion phenomena: Chemical and non-chemical bonds operate. Adequate experimental techniques for measuring adhesion strength are still lacking.

When considering adhesion, one must consider elastic and non-elastic deformation that may take place at the point of attachment. The short-range forces could be strong, e.g. primary bonds, or intermediate, e.g. hydrogen and charge-transfer bonds.

In 1934, Deryaguin [10, 11] considered the force of adhesion F in terms of the free energy of separation of two surfaces, $G(h_\infty) - G(h_0)$, from a distance h_0 to infinite separation distance h_∞. For the simple case of parallel plates,

$$-\int_{h_0}^{\infty} F\,dh = G(h_\infty) - (h_0) \tag{6}$$

F is made of three contributions:

$$F = F_m + F_c + F_e \tag{7}$$

where F_m is the molecular component and consists of two components, an elastic deformation component F_S and a surface energy component F_H:

$$F_m = F_S + F_H \tag{8}$$

F_c is the component that depends on prior electrification and F_e is the electrical double layer contribution.

When a sphere adheres to a plane surface, elastic deformation occurs and one can distinguish the radius of the adhesive area r_0. A schematic representation of elastic deformation is shown in Figure 5.10.

Usually $r_0/R \ll 1$, where R is the particle radius. The adhesive area can be calculated from a knowledge of the time dependence of the modulus of the sphere and the time dependence of the hardness of the plate.

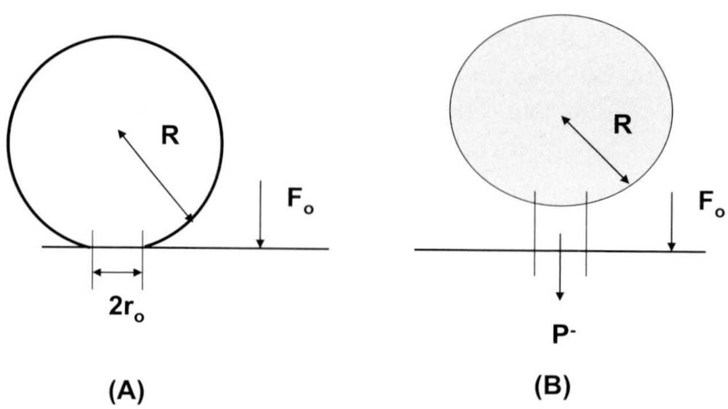

Figure 5.10 Elastic deformation on adhesion of a sphere to a plane surface.

5.2.1
Surface Energy Approach to Adhesion

Two approaches may be applied to consider the process of adhesion: the Fox and Zisman critical surface tension approach and Neuman's equation of state approach.

5.2.1.1 Fox and Zisman Critical Surface Tension Approach

This approach was initially used to obtain the critical surface tension of wetting of liquids on solid substrates. Fox and Zisman [12] found that a plot of cos θ (where θ is the contact angle of a liquid drop on the substrate) versus γ_{LV} (the liquid surface tension) for a number of related liquids gives a straight line which, when extrapolated to cos $\theta = 1$, gives the critical surface tension of wetting γ_c. This is shown in Figure 5.11 for a number of solids.

A liquid with $\gamma_{LV} < \gamma_c$ will give complete wetting of the substrate. Surfaces with high $\gamma_c > 40$ mN m^{-1} and small slopes are high-energy surfaces (e.g. glass and cellulose). Surfaces with low very γ_c (<22 mN m^{-1}) and high slope are low-energy surfaces, e.g. Teflon. Hydrocarbon surfaces such as Vaseline produce intermediate values ($\gamma_c \approx 30$ mN m^{-1}).

The above approach could also be applied to the adhesion of 'soft' particles to solid substrates. One can define three surface tensions: γ_{PL} (particle/liquid), γ_{PS} (particle/surface) and γ_{SL} (solid/liquid).

The Helmholtz free energy of adhesion ΔF is given by the following expression:

$$\Delta F = (\gamma_{PS} - \gamma_{PL} - \gamma_{SL})\pi r_0^2 \qquad (9)$$

For adhesion to occur, ΔF should be negative. If ΔF is positive, no adhesion occurs.

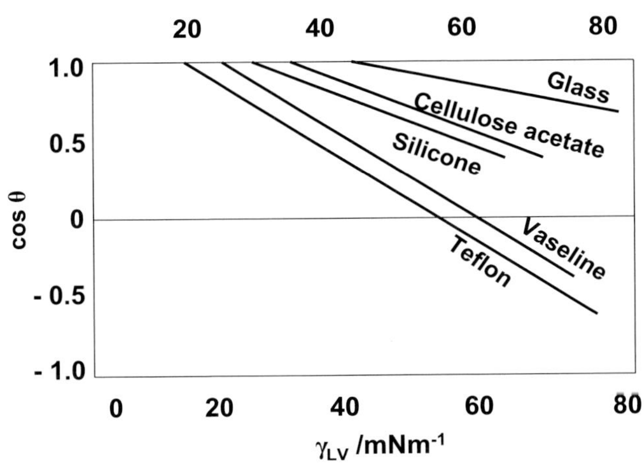

Figure 5.11 Cos θ versus γ_{LV} for a number of substrates.

5.2.1.2 Neuman's Equation of State Approach

Neuman [13] simplified the analysis by using a simple equation of state approach. He showed that a plot of γ_{LV} cos θ versus γ_{LV} gives a smooth curve, as illustrated in Figure 5.12. This simple analysis allows one to obtain γ_S from a single contact angle measurement.

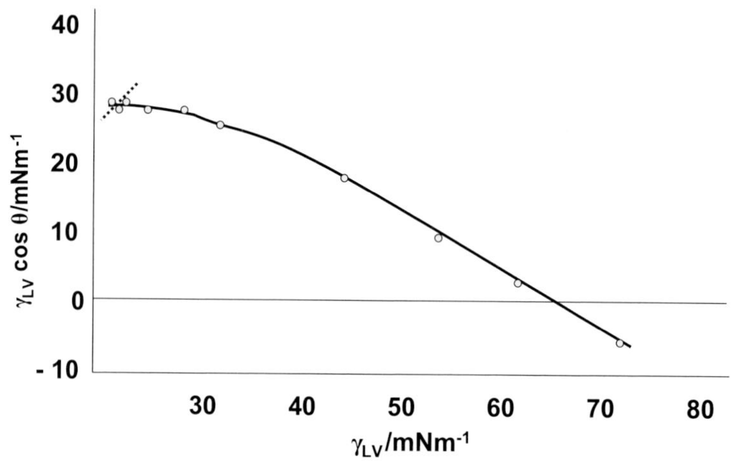

Figure 5.12 Plot of γ_{LV} cos θ versus γ_{LV}.

5.2.2
Experimental Methods for Measurement of Particle–Surface Adhesion

One has to measure the force required to remove the particle from the substrate, and methods that can be used are outlined below.

5.2.2.1 Centrifugal Method (Krupp, 1967)

The centrifugal force required to remove a particle is given by the expression [14]

$$F_c = \frac{4}{3}\pi a^3(\rho_p - \rho_w)(\omega^2 x + g) \tag{10}$$

where a is the particle radius, ρ_p is the particle density, ρ_w is the density of water, ω is the centrifuge speed, x is the distance from the rotor and g is the acceleration due to gravity.

To remove small particles from substrates, one needs to apply very high g values (as high as 10^7). Hence this method is of little practical application.

5.2.2.2 Hydrodynamic Method (Visser, 1970)

A rotating cylinder apparatus is used and, after the particles have been deposited on the inner cylinder, the speed is increased [15]. The percentage detachment is measured (microscopically) as a function of the speed of rotation. The hydrodynamic force required to remove 50% of the particles is taken as a measure of the force of adhesion.

The above method was applied by Tadros [16] in 1980 to measure the force of adhesion of polystyrene latex on polyethylene. The results are shown in Figure 5.13, which shows the percentage of particles removed versus the speed of rotation of the cylinder.

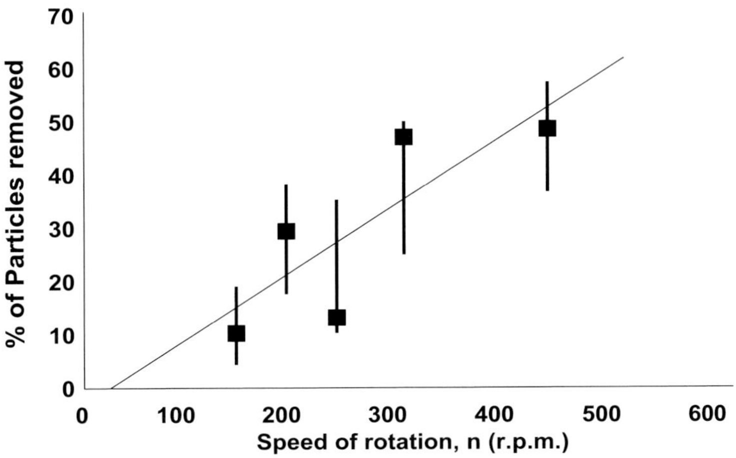

Figure 5.13 Percentage of particles removed versus speed of rotation.

From a knowledge of the hydrodynamic force required for particle removal, one could calculate the force of adhesion. The force of adhesion could be compared with the attractive force calculated from the Hamaker equation. The results showed that the force of adhesion was about two orders of magnitude lower than the theoretical value calculated from the van der Waals attraction.

It was concluded from the above results that the latex particles are not perfectly smooth ('hairy' surface) and hence not in intimate contact with the surface.

References

1 H.C. Hamaker, *Physica (Utrecht)*, *4*, 1058 (1937).

2 G. Gouy, *J. Phys.*, (4), *9*, 457 (1910); *Ann. Phys.*, *7* (9), 129 (1917); D.L. Chapman, *Philos. Mag.*, *25* (6), 475 (1913).

3 O. Stern, *Z. Elektrochem.*, *30*, 508 (1924).

4 B.V. Deryaguin and L. Landau, *Acta Physicochim. URSS*, *14*, 633 (1941).

5 E.J.W. Verwey and J.Th.G. Overbeek, *Theory of Stability of Lyophobic Colloids*, Elsevier, Amsterdam (1948).

6 M. Hull and S.A. Kitchener, *Trans. Faraday Soc.*, *65*, 3039 (1969).

7 Th.F. Tadros, unpublished results.

8 J.R. Huntsberger, The mechanism of adhesion, in *Treatise on Adhesion and Adhesives*, ed. R.L. Kilpatrik, Marcel Dekker, New York, Chapter 4 (1967).

9 Th.F. Tadros, Particle–surface adhesion, in *Microbial Adhesion to Surfaces*, ed. R.C.W. Berkeley *et al.*, Ellis Horwood, Chichester, Chapter 5 (1980).

10 B.V. Deryaguin, *Kolloid Z.*, *69*, 155 (1934).

11 B.V. Deryaguin and V.P. Smilga, in *Proceedings of the 3rd International Congress on Surface Activity*, Cologne, Vol. II, December 1960, p. 349 (1960).

12 H.W. Fox and W.A. Zisman, *J. Colloid Sci.*, *109* (7), 428 (1952).

13 A.W. Neuman, *Adv. Colloid Interface Sci.*, *4*, 105 (1974).

14 H. Krupp, *Adv. Colloid Interface Sci.*, *1*, 111 (1967).

15 J. Visser, *J. Colloid Interface Sci.*, *34*, 26 (1970).

16 Th.F. Tadros, unpublished results.

6
Basic Principles of Rheology

One of the most important characteristics of any paint formulation is its flow properties (rheology), which are very important both for its long-term physical stability and for its application. In the latter case, it is important to control the rheology for ease of application and production of the required paint film. Two types of rheological measurements must be applied, namely steady-state and viscoelastic techniques. The basic principles of each type of measurement are described below together with the experimental methods that are applied.

6.1
Principles of Steady-state Measurements

In this case, the dispersion, which is placed in the gap between concentric cylinders, cone and plate or parallel plates, is subjected to a constant shear rate until a steady state is reached, after which the stress remains constant [1–4]. The measurement allows one to obtain the stress and viscosity as a function of shear rates. Before describing these principles, it is useful to define the stress, strain, strain (shear) rate and viscosity. The laws of elasticity (Hooke's law) and viscosity (Newton's law) are briefly described. The reversible time effects of viscosity (thixotropy and dilatancy) are also discussed.

Stress (σ): The deformation produced by a force acting on a material depends on the magnitude of the force per unit area (stress); σ is given in N m^{-2} or Pa (SI units, kg m s), and it can also be expressed in dyn cm^{-2} (c.g.s. units). Note that 1 Pa = 10 dyn cm^{-2}.

Strain (γ): A stress applied to a material results in its deformation. This is illustrated in Figure 6.1, which shows a schematic representation of deformation of a material represented by ABC when a stress σ is applied. The material acquires a new shape A′B′C′. Note that AB ≠ A′B′, AC ≠ A′C′ and ABC ≠ A′B′C′.

In simple shear strain, successive layers of material move in their own planes, relative to a reference layer. The displacement of a layer is proportional to its distance from the reference layer shown schematically in Figure 6.2 for two layers, represented by a reference layer (similar to a deck of cards that can only be deformed sideways).

Colloids and Interface Science Series, Vol. 6
Colloids in Paints. Tharwat F. Tadros

ISBN: 978-3-527-31466-9

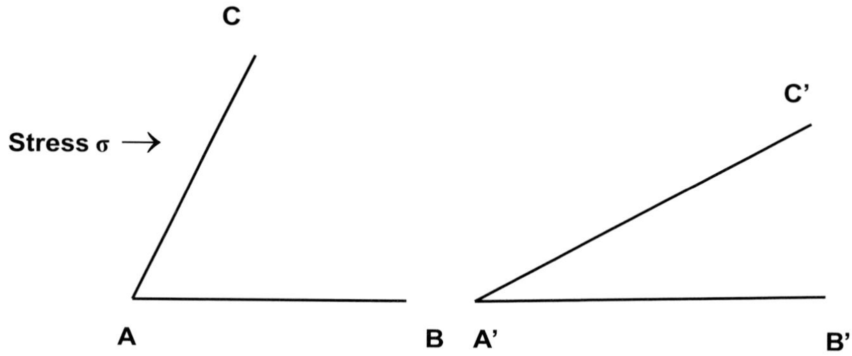

Figure 6.1 Schematic representation of deformation on application of stress.

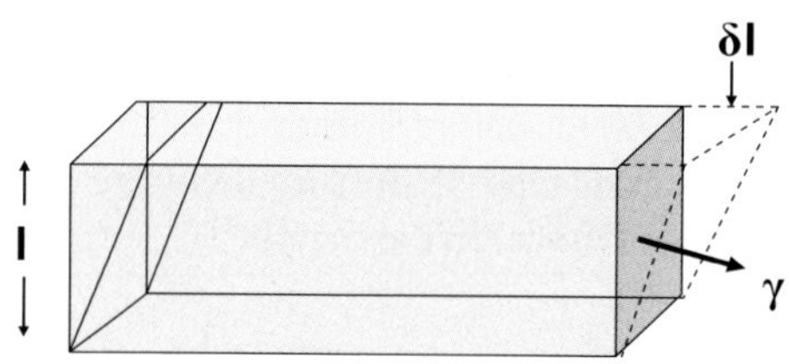

Figure 6.2 Deformation of a rectangular element in simple shear.

$$\text{Strain} = \frac{\mathrm{d}l}{l} \tag{1}$$

$$\tan\gamma = \frac{\mathrm{d}l}{l} \tag{2}$$

When γ is very small, $\tan\gamma \approx \gamma$. The shear strain is simply the angle of shear in radians.

6.1.1
Strain Rate or Shear Rate

The shear rate is simply given by the change of strain with time; it has the unit s^{-1}.

$$\dot{\gamma} = \frac{\mathrm{d}\gamma}{\mathrm{d}t} \tag{3}$$

When the flow is linear, a layer at a distance l from the reference layer moving with a velocity v will travel a distance $\mathrm{d}l = v\,\mathrm{d}t$, so that

$$\dot{\gamma} = \frac{\mathrm{d}}{\mathrm{d}t}\left(\frac{\mathrm{d}l}{l}\right) = \frac{v}{l}; \quad \text{units } \frac{\mathrm{ms}^{-1}}{\mathrm{m}} = \mathrm{s}^{-1} \tag{4}$$

6.1.2
Types of Rheological Behavior in Simple Shear

Consider a layer of material placed between two large parallel plates (edge effects neglected) to which it adheres and a shear stress σ is applied via the plates. If the stress is removed, the strain may or may not return to zero, and this allows one to distinguish the flow behavior. Strain does not return to zero: flow has occurred and the material is considered as a liquid. If flow occurs even for a vanishingly small stress, the material is still considered as a liquid. For inelastic solids, no strain recovery occurs; elastic liquids, partial recovery occurs, e.g. adhesives; ideally elastic, deformation and recovery are instantaneous; slow recovery, viscoelastic, e.g. polymer solutions, suspensions and emulsions.

6.1.3
Models for Flow Behavior

6.1.3.1 Law of Elasticity (Hooke's Model)

The stress (force per unit area) is proportional to the relative deformation (strain). A plot of stress σ versus strain γ gives a straight line (at small applied strains) with a slope equal to the shear modulus G':

$$\sigma = G'\gamma \tag{5}$$

Figure 6.3 shows the relationship for an elastic solid. A material that obeys Hooke's law is described as a Hookean solid. In this case, G' is independent of the applied stress.

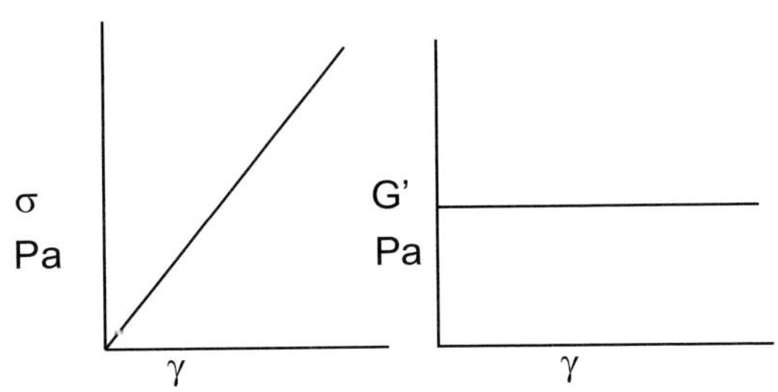

Figure 6.3 Hooke's law of elasticity.

6.1.3.2 Newton's Law of Viscosity

The resistance which arises from the lack of slipperiness of parts of a liquid (other things being equal) is proportional to the velocity with which the parts of the liquid are separated from one another. The strain rate $\dot{\gamma}$ is a single-valued function of the stress σ:

$$\dot{\gamma} = f(\sigma) \tag{6}$$

In other words, the stress σ versus shear rate $\dot{\gamma}$ is linear and the slope is equal to the viscosity η (Pa s):

$$\sigma = \eta\dot{\gamma} \tag{7}$$

Plots of σ and η versus shear rate are given in Figure 6.4. A material that obeys Newton's law is described as Newtonian.

The viscosity η is given by the ratio of shear stress to shear rate:

$$\eta = \frac{\sigma}{\dot{\gamma}}; \quad \text{units} \ \frac{\text{N m}^{-2}}{\text{s}^{-1}} = \text{N m}^{-2}\ \text{s} = \text{Pa s} \tag{8}$$

or

$$\eta\text{: units} = \text{dyn cm}^{-2}\ \text{s} = \text{poise (P)} \tag{9}$$

Viscosity of water at ~20 °C:
1 Pa s = 10 P
10^{-3} Pa s = 1 mPa s
10^{-2} P = 1 cP
1 mPa s = 1 cP.

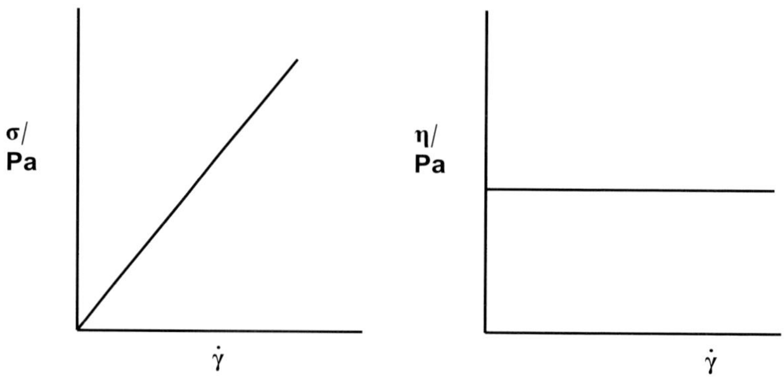

Figure 6.4 Newton's law.

6.1.3.3 **Non-Newtonian Flow**

Most dispersions, particularly those with high volume fraction and/or containing rheology modifiers, do not obey Newton's law. This can be clearly shown from plots of shear stress σ versus shear rate, as illustrated in Figure 6.5. Five different flow curves can be identified: (a) Newtonian; (b) Bingham plastic; (c) pseudoplastic (shear thinning); (d) dilatant (shear thickening); and (e) yield stress and shear thinning. The variation of viscosity with shear rate for the above five systems is shown in Figure 6.6. Apart from the Newtonian flow (a), all other systems show a change in viscosity with applied shear rate.

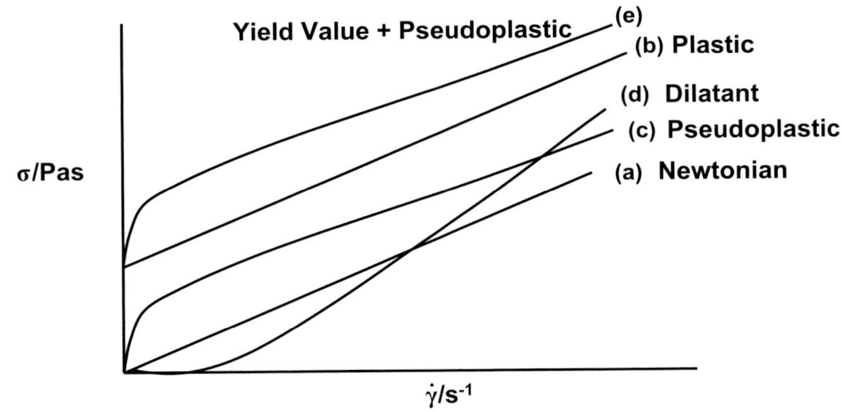

Figure 6.5 Viscosity–shear rate relationship.

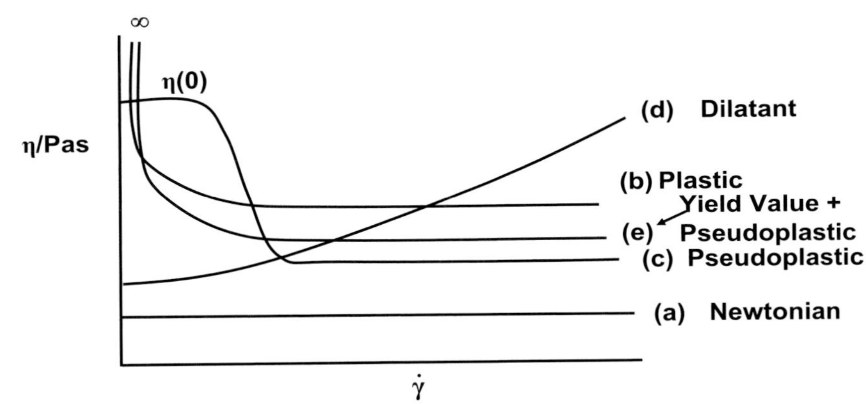

Figure 6.6 Flow curves for various systems.

The flow curves shown can be fitted to various models that allow one to obtain the required rheological parameters. Several models have been presented and these are described below. It should be mentioned that these models are semi-empirical in nature and the model that gives the best fit to the experimental data (with a very good correlation coefficient that is >0.99) is used.

6.1.4 Rheological Models for Analysis of Flow Curves

6.1.4.1 Newtonian Systems

$$\sigma = \eta\dot{\gamma} \tag{10}$$

where η is independent of the applied shear rate, e.g. simple liquids and very dilute dispersions.

6.1.4.2 Bingham Plastic Systems [5]

$$\sigma = \sigma_\beta + \eta_{pl}\dot{\gamma} \tag{11}$$

The system shows a (dynamic) yield stress σ_β that can be obtained by extrapolation to zero shear rate. Clearly, at and below σ_β the viscosity $\eta \to \infty$. The slope of the linear curve gives the plastic viscosity η_{pl}. Some systems such as clay suspensions may show a yield stress above a certain clay concentration.

6.1.4.3 Pseudoplastic (Shear Thinning) System

In this case, the system does not show a yield value. It shows a limiting viscosity $\eta(0)$ at low shear rates (that is referred to as residual or zero shear viscosity). The flow curve can be fitted to a power law fluid model (Ostwald de Waele):

$$\sigma = k\dot{\gamma}^n \tag{12}$$

where k is the consistency index and n is the shear thinning index; $n < 1$.

By fitting the experimental data to equation 12, one can obtain k and n. The viscosity at a given shear rate can be calculated as

$$\eta = \frac{\sigma}{\dot{\gamma}} = \frac{k\dot{\gamma}^n}{\dot{\gamma}} = k\dot{\gamma}^{n-1} \tag{13}$$

6.1.4.4 Dilatant (Shear Thickening) System

In this case the viscosity increases with applied shear rate and the flow curve can be fitted with the power law (equation 13), but in this case $n > 1$. Several systems can show shear thickening such as wet sand, corn starch dispersed in milk and some poly(vinyl chloride) sols. Shear thickening can be illustrated when one walks on wet sand, where some water is 'squeezed out' and the sand appears dry. The deformation applied by one's foot causes rearrangement of the close-packed structure produced by the water motion. This process is accompanied by a volume increase (hence the term dilatancy) as a result of 'sucking in' of the water. The process leads to a rapid increase in the viscosity.

6.1.4.5 Herschel–Bulkley General Model [6]

Many systems show a dynamic yield value followed by a shear thinning behavior. The flow curve can be analyzed using the Herschel–Bulkley equation:

$$\sigma = \sigma_\beta + k\dot{\gamma}^n \tag{14}$$

When $\sigma_\beta = 0$, equation 14 reduces to the power fluid model. When $n = 1$, equation 14 reduces to the Bingham model. When $\sigma_\beta = 0$ and $n = 1$, equation 14 becomes the Newtonian equation.

The Herschel–Bulkley equation fits most flow curves with a good correlation coefficient, hence it is the most widely used model.

Several other models have been suggested, of which the following are worth mentioning:

6.1.4.6 The Casson Equation [7]

This is a semiempirical linear parameter model that has been applied to fit the flow curves of many paints and printing ink formulations:

$$\sigma^{\frac{1}{2}} = \sigma_C^{\frac{1}{2}} + \eta_C^{\frac{1}{2}}\dot{\gamma}^{\frac{1}{2}} \tag{15}$$

Hence a plot of $\sigma^{\frac{1}{2}}$ versus $\dot{\gamma}^{\frac{1}{2}}$ should give a straight line from which σ_C and η_C can be calculated from the intercept and slope of the line.

One should be careful when using the Casson equation since straight lines are obtained from the results only above a certain shear rate.

6.1.4.7 The Cross Equation [8]

This can be used to analyze the flow curve of shear thinning systems that show a limiting viscosity $\eta(0)$ in the low shear rate regime and another limiting viscosity $\eta(\infty)$ in the high shear rate regime. These two regimes are separated by a shear thinning behavior as shown schematically in Figure 6.7.

$$\frac{\eta - \eta(\infty)}{\eta(0) - \eta(\infty)} = \frac{1}{1 + \alpha\dot{\gamma}^n} \tag{16}$$

where α is the consistency index and n is the shear thinning index.

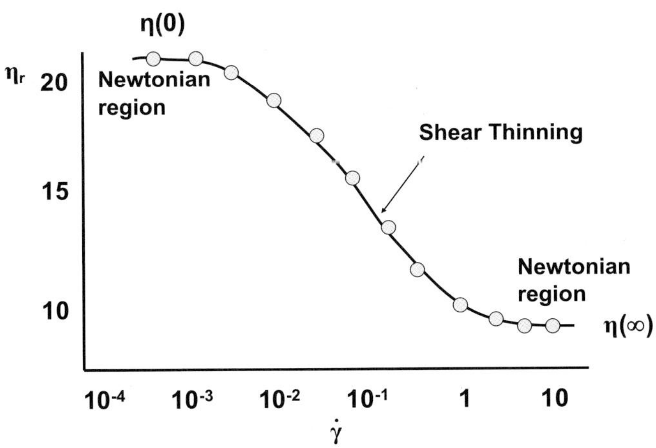

Figure 6.7 Viscosity versus shear rate for shear thinning system.

6.1.5
Time Effects During Flow – Thixotropy and Negative (or Anti-) Thixotropy

When a shear rate is applied to a non-Newtonian system, the resulting stress may not be achieved simultaneously. (i) The molecules or particles will undergo spatial rearrangement to follow the applied flow field. (ii) The structure of the system may change: breaking of weak bonds; aligning of irregularly shaped particles; collision of particles to form aggregates

The above changes are accompanied by a decrease or increase in viscosity with time at any given shear rate. These changes are referred to as thixotropy (if the viscosity decreases with time) or negative thixotropy or antithixotropy (if the viscosity increases with time).

Thixotropy refers to the reversible time-dependent decease in viscosity. When the system is sheared for some time, the viscosity decreases, but when the shear is stopped (the system is left to rest), the viscosity of the system is restored. Practical examples for systems that show thixotropy are paint formulations (sometimes referred to as thixotropic paints), tomato ketchup and some hand creams and lotions.

Negative thixotropy or antithixotropy: when the system is sheared for some time, the viscosity increases, but when the shear is stopped (the system is left to rest), the viscosity decreases. A practical example of the above phenomenon is corn starch suspended in milk.

The most common procedure for studying thixotropy is to apply a sequence of shear rate–shear stress cycles within well-defined time scales. This is illustrated in Figure 6.8. Cycle 1 is carried out within a short time scale: the shear rate is increased from, say, 0 to 500 s^{-1} in 30 s and the stress is measured simultaneously. At the end, the shear rate is decreased from 500 to 0 s^{-1} in 30 s. The sys-

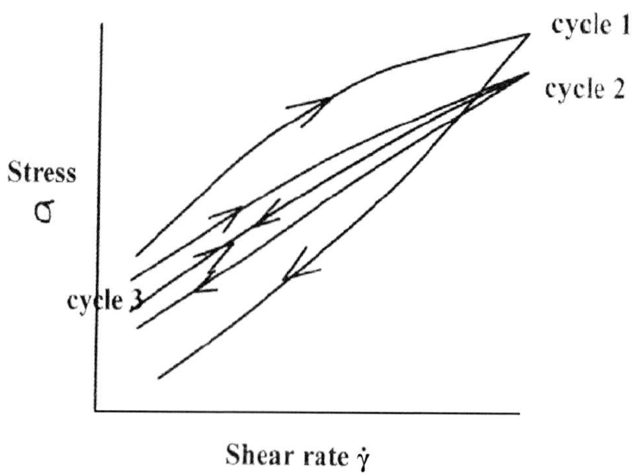

Figure 6.8 Thixotropic loops.

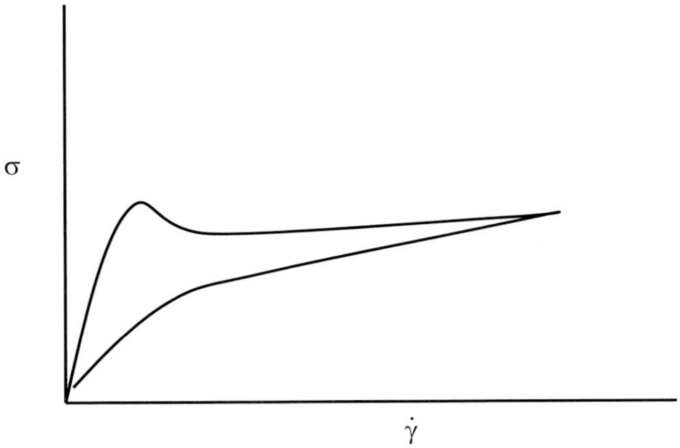

Figure 6.9 Elastic overshoot.

tem does not have enough time to recover and it shows hysteresis (the down curve is below the up curve). This is referred to as a thixotropic loop. Cycle 2: the above experiment is now carried out over a longer time, say 120 s for the up curve and 120 s for the down curve. The thixotropic loop is smaller in area. Cycle 3 is carried out over a sufficiently long time for recovery to be established and the loop closes.

The area under the loop may be taken as a qualitative measure of thixotropy to compare various systems. This analysis may not be straightforward since many systems may also show an elastic overshoot, as indicated in Figure 6.9. Simple non-linear relaxation behavior may show a complex behavior of the stress with time. This is illustrated in Figure 6.10, which shows the change in stress with

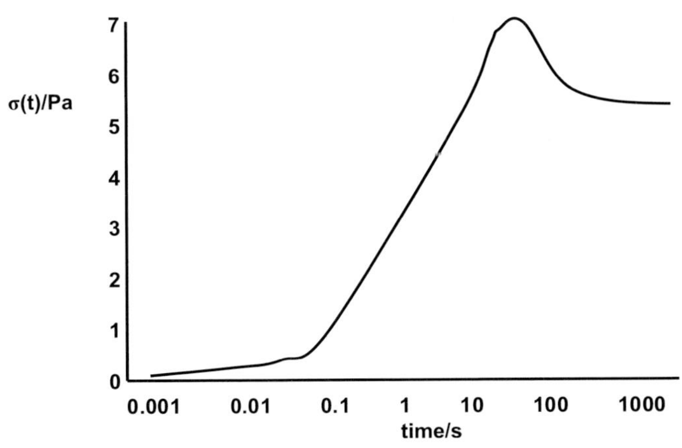

Figure 6.10 Variation of stress with time at a strain $\gamma = 1$.

time at a shear strain $\gamma = 1$. The peak observed is referred to as elastic overshoot. For accurate investigation of thixotropy, it is preferable to use viscoelastic measurements, as will be discussed in the next chapter.

6.1.6 Rheopexy

Rheopexy is the increase in thixotropy recovery rate when gentle oscillation is applied to the system. Rheopexy is sometimes qualitatively described as an 'increase in viscosity by touch'. It should not be confused with the term antithixotropy, which is used to describe time-dependent shear thickening.

The best way to investigate rheopexy is to follow the modulus G as a function of time t on application of strains with increasing amplitudes (see the next section). With a rheopectic system, G increases more rapidly with time at higher amplitudes compared with lower amplitudes. Care must be taken when characterizing the thixotropic recovery to avoid enhancement of the rate.

6.1.7 Measurement of Viscosity as a Function of Shear Rate – Steady-state Regime

Measurement of viscosity as a function of shear rate is achieved when a steady state is reached. It is important to consider the shear rates involved in various industrial processes. These are summarized in Table 6.1. Table 6.2 gives a summary of the viscosities of various materials given on a logarithmic scale to cover the whole range.

Steady-state measurements can be applied to measure the viscosity as a function of shear rate in the range 10^{-1}–10^{3} s^{-1}. Measurement at very low shear rates (the range that is important for the prediction of creaming or sedimentation) can only be achieved using constant stress (creep) measurements. Shear rates above 10^{3} s^{-1} require special instruments, e.g. using capillary viscometry.

Table 6.1 Shear rates of various processes.

Process	Shear rate $\dot{\gamma}$/s^{-1}	Application
Sedimentation	10^{-6}–10^{-4}	Paints, agrochemicals
Leveling	10^{-2}–10^{-1}	Paints, printing inks
Extrusion	10^{0}–10^{2}	Polymer processing
Stirring, mixing	10^{1}–10^{2}	Manufacturing
Pipeline flow	10^{0}–10^{3}	Pumping
Ball milling	10^{3}–10^{5}	Paints, inks
Brushing, spraying	10^{3}–10^{4}	Coatings
Topical rubbing	10^{4}–10^{5}	Cosmetics
High-speed coating	10^{3}–10^{7}	Paper coating

Table 6.2 Viscosities of some typical materials.

Viscosity/Pa s	Material
10^6	Polymer melts
10^4	
10^2	
10^0	Glycerol
10^{-2}	Ethylene glycol
10^{-4}	Water
10^{-6}	Air

6.1.8
Capillary Viscometers

These are essential for calibration purposes (using standard liquids), measuring Newtonian systems such as dilute polymer solutions, dilute suspensions and emulsions, etc. The most widely used capillary viscometer is the Ostwald type shown schematically in Figure 6.11. A variant of the Ostwald viscometer is the Cannon–Fenske type (Figure 6.11), which is more convenient to use. In capillary viscometry, one measures the volumetric flow Q ($m^3\ s^{-1}$) and the viscosity η is calculated using Poiseuille's equation:

$$\eta = \frac{\pi R^4 p}{8QL} \tag{17}$$

where R is the tube radius with length L and p is the pressure drop $= h\rho g$ (where h is the liquid height with density ρ and g is the acceleration due to gravity).

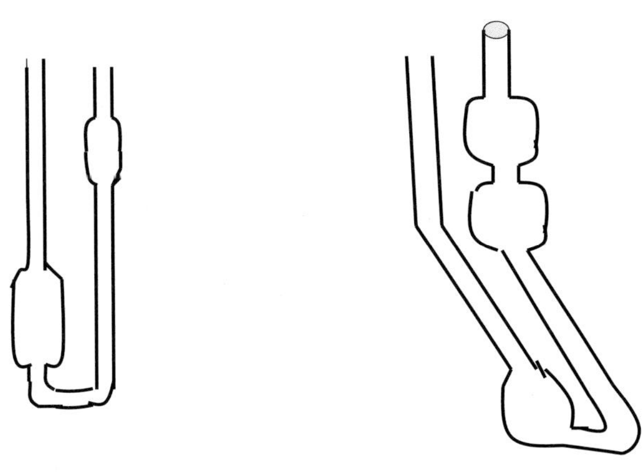

Ostwald Viscometer **Cannon-Fenske Viscometer**

Figure 6.11 Schematic diagrams of Ostwald and Cannon–Fenske viscometers.

One usually compares the viscosity of the liquid in question η_2 with that of a liquid with known viscosity η_1. In this way, one can measure the flow rates of the two liquids using the same viscometer with a bulb of volume V (the flow rate is simply given by V divided by the time taken for the liquid to flow between the two marks on the viscometer t_1 and t_2 for the two liquids).

Using equation 17 one simply obtains

$$\frac{\eta_1}{\eta_2} = \frac{t_1 \rho_1}{t_2 \rho_2} \tag{18}$$

Times t_1 and t_2 are simply measured using a stop-watch (for automatic viscometers two fiber optics are used). Accurate temperature control is necessary (± 0.01 °C). The flow time t must also be measured with an accuracy of ± 0.01 s.

6.1.9
Measurement of Intrinsic Viscosity of Polymers

Capillary viscometers are useful for the measurement of the intrinsic viscosity $[\eta]$ of polymers, which can be used to obtain the molecular weight and solvation of the polymer chains. One measures the relative viscosity η_r as a function of polymer concentration C (in the range 0.01–0.1%):

$$\eta_r = \frac{\eta_s}{\eta_0} \tag{19}$$

where η_s is the viscosity of the polymer solution and η_0 is that of the solvent.

From η_r, one can obtain the specific viscosity η_{sp}:

$$\eta_{sp} = \eta_r - 1 \tag{20}$$

From η_{red}, one can obtain the reduced viscosity η_{red}:

$$\eta_{red} = \frac{\eta_{sp}}{C} \tag{21}$$

A plot of η_{sp} versus C gives a straight line that can be extrapolated to $C = 0$ to obtain the intrinsic viscosity $[\eta]$. This is illustrated in Figure 6.12.

From $[\eta]$, one can obtain the molecular weight M using the Mark–Houwink equation:

$$[\eta] = KM^{\alpha} \tag{22}$$

where K and α are constants for a particular polymer and solvent (values for many polymer–solvent systems are tabulated in the *Polymer Handbook*). The constant α is related to the solvency of the medium for the polymer chain. In a good solvent

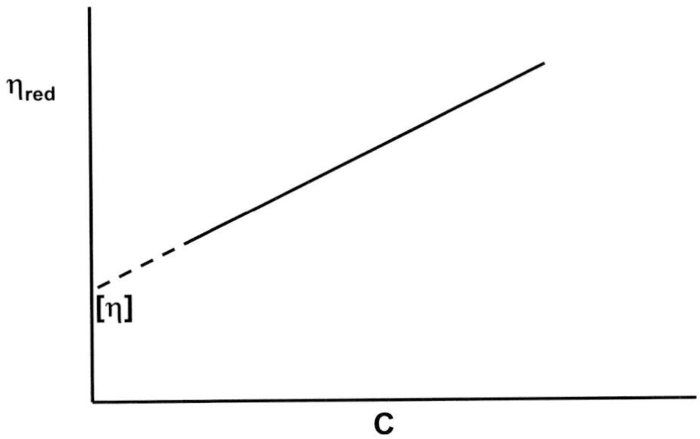

Figure 6.12 Measurement of intrinsic viscosity of polymers.

$\alpha > 0.5$; it has values in the range 0.5–0.8. The higher the value of α, the better the solvent is for the chain.

6.1.10
Capillary Rheometry for Non-Newtonians

Capillary rheometers can be used for non-Newtonian systems, particularly when high shear rates need to be applied. The flow rate Q is measured as a function of pressure drop p. The stress at the wall σ_W can be calculated as

$$\sigma_W = \frac{Rp}{2L} \tag{23}$$

The apparent viscosity η_a is then calculated at each pressure drop using the Poiseuille equation 17. Graphs are then plotted of log η_a versus log σ_W. The slope m at each data point is then calculated:

$$m = \frac{\mathrm{d} \log \eta_a}{\mathrm{d} \log \sigma_W} \tag{24}$$

The viscosity at the wall $\eta(\sigma_W)$ is calculated using the Weissenberg equation;

$$\eta(\sigma_W) = \frac{\eta_a}{1 - \frac{m}{4}} \tag{25}$$

The shear rate at the wall can also be calculated:

$$\dot{\gamma}(\sigma_W) = \frac{\sigma_W}{\eta(\ )\sigma_W} = \frac{\sigma_W}{\eta_a}\left(1 - \frac{m}{4}\right) \tag{26}$$

6.1.11
Rotational Viscometers

Rotational viscometers have advantages over tube viscometers: (i) the sample can be sheared for any length of time and (ii) the shear rate is approximately constant (particularly with cone and plate geometry).

Three main geometries are commonly used: (i) concentric cylinder (Couette), (ii) cone and plate and (iii) parallel plate. The concentric cylinder geometry is the most convenient due to its high sensitivity (high surface area) and minimum sample evaporation. The shear rate is not uniform in the gap. The cone and plate geometry is the most accurate (uniform shear rate in the gap). The parallel plate geometry has the advantage of the possibility of changing the gap width (but the shear rate is not uniform).

6.1.12
Concentric Cylinder Viscometer

This viscometer consists of an inner (bob) and outer (cup) concentric cylinders. Aschematic diagram is shown in Figure 6.13. The bob may contain a concave base which is designed to entrap air in order to reduce the drag contribution from the base.

Let us first consider a Newtonian liquid that is placed in the gap between the two cylinders with radii R_1 (inner) and R_2 (outer) and length L. The stress σ is simply the force F per unit area A, the force being the ratio of the torque M to the radius R and the area being $2\pi RL$:

$$\sigma = \frac{F}{A} = \frac{M/R}{2\pi RL} = \frac{M}{2\pi R^2 L} \qquad (27)$$

One can define a stress at the inner cylinder σ_1 and at the outer cylinder σ_2:

$$\sigma_1 = \frac{M}{2\pi {R_1}^2 L} \qquad (28)$$

$$\sigma_2 = \frac{M}{2\pi {R_2}^2 L} \qquad (29)$$

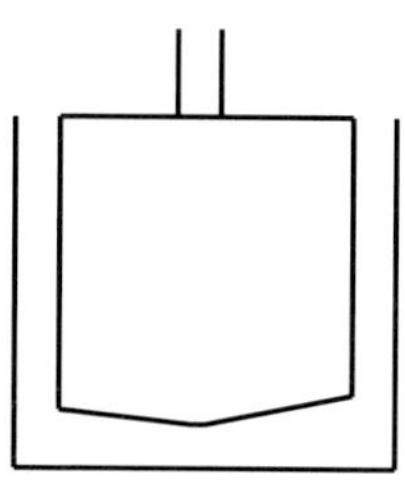

Figure 6.13 Schematic diagram of a concentric cylinder.

The shear rate can be calculated by assuming that the fluid elements move in circles about the common axis if the cylinders with an angular velocity ω, which is a function of radius r from the center to the position in the gap (where the stress is σ):

$$\dot{\gamma} = r\frac{\mathrm{d}\omega}{\mathrm{d}r} \tag{30}$$

$$\mathrm{d}\omega = \frac{\dot{\gamma}}{r}\mathrm{d}r = \frac{M\,\mathrm{d}r}{2\pi r^3 L\eta} \tag{31}$$

To calculate the angular velocity Ω of the outer cylinder, we have to sum all the contributions from the inner to the outer cylinder:

$$\Omega = \int_{R_1}^{R_2} \mathrm{d}\omega = \frac{M}{2\pi L\eta}\int_{R_1}^{R_2}\frac{\mathrm{d}r}{r^3} = \frac{M}{4\pi L\eta}\left(\frac{1}{{R_1}^2} - \frac{1}{{R_2}^2}\right) \tag{32}$$

Equation 32, referred to as the Margules equation, shows that a plot of Ω versus M should give a straight line passing through the origin, from which η can be obtained from the slope.

6.1.13
Non-Newtonians

6.1.13.1 Shear Thinning or Pseudoplastic

This could follow a power law:

$$\sigma = k\dot{\gamma}^n \tag{33}$$

In this case, a plot of Ω versus M is not linear; a log–log plot is linear and the slope gives $N = 1/n$.

The stress σ_1 at the inner cylinder is calculated for each data point:

$$\sigma_1 = \frac{M}{2\pi {R_1}^2 L} \tag{34}$$

and the shear rate at the inner cylinder is calculated as

$$\dot{\gamma}_1 = \frac{2N\Omega}{1 - s^{-2N}} \tag{35}$$

where $s = R_2/R_1$.

6.1.13.2 Bingham Plastic

Here

$$\sigma = \sigma_\beta + \eta_{pl}\dot{\gamma} \tag{36}$$

In this case, a plot of Ω versus M is linear with an intercept that gives the yield value and a slope that gives the plastic viscosity:

$$\Omega = \frac{M}{4\pi\eta_{pl}}\left(\frac{1}{R_1{}^2} = \frac{1}{R_2{}^2}\right) - \frac{\sigma_\beta}{\eta_{pl}}\ln\frac{R_2}{R_1} \tag{37}$$

It is assumed that all the material is flowing, i.e. the shear stress at the outer cylinder is higher than the yield value:

$$\frac{M}{2\pi R_2{}^2 L} \geq \sigma_\beta \tag{38}$$

6.1.14 Major Precautions with Concentric Cylinder Viscometers

6.1.14.1 Shear Rate Calculations

The shear rate in the gap is not uniform and it varies according to the position r in the gap. Using the Margules equation 32, one can calculate the shear at any distance r in the gap. The stress σ at r is given by

$$\sigma = \frac{F}{A} = \frac{M/r}{2\pi r L} = \frac{M}{2\pi r^2 L} \tag{39}$$

Combining equations 39 and 32, the shear rate at r is given by

$$\dot{\gamma} = \frac{\sigma}{\eta} = \frac{2\Omega}{r^2}\left(\frac{1}{R_1{}^2} - \frac{1}{R_2{}^2}\right)^{-1} \tag{40}$$

One can calculate the shear rate at the inner and outer cylinders as follows:
At $r = R_1$:

$$\dot{\gamma}_1 = 2\Omega\frac{R_2{}^2}{R_2{}^2 - R_1{}^2} \tag{41}$$

At $r = R_2$:

$$\dot{\gamma}_2 = 2\Omega\frac{R_1{}^2}{R_2{}^2 - R_1{}^2} \tag{42}$$

$$\langle\dot{\gamma}\rangle = \Omega\frac{{R_2}^2 + {R_1}^2}{{R_2}^2 - {R_1}^2} \tag{43}$$

which is the mean value.

Simple calculations show that if the gap is 5% of the outer cylinder radius, the shear rate varies by 5% across the gap. If $R_2 - R_1$ is made sufficiently small, i.e. much smaller than R_1, an average shear stress and shear rate can be calculated:

$$\dot{\gamma}_{\mathrm{av}} = \frac{\Omega R_2}{R_2 - R_1} \tag{44}$$

$$\sigma_{\mathrm{av}} = \frac{M}{2\pi {R_{\mathrm{a}}}^2} \tag{45}$$

The above equations are only valid if $R_2 - R_1 \approx 1\%$ of R_1. This is not possible, for two reasons: (i) it is difficult to machine cylinders with such a high tolerance and (ii) the gap width must be much larger ($\gg 10\times$) the largest particle or droplet size (which may exceed 10 μm). Gap widths in the range 500–1000 μm are usually used and one can refer to the shear rate at the outer or inner cylinder or the mean value. If a constant shear rate in the gap is required, one should use a cone and plate geometry as discussed below.

6.1.14.2 Wall Slip and Sample Evaporation During Measurement

Many suspensions and emulsions that are concentrated and/or 'structured' or 'weakly flocculated' (to avoid separation) may cause wall slip. The depletion of particles near the wall can lead to erroneously low viscosity and low yield values. The slip can be prevented by roughening the inside surfaces of the cylinders. However, one should be careful in using rough surfaces that may cause fluctuation in the shear rate. Alternatively, one may use different gap widths and extrapolate the results to a large gap width (assuming that the slip layer is independent of gap width). A better approach to avoid wall slip is to use a vane rheometer, as discussed below. This is particularly useful for highly concentrated systems such as ceramic pastes. Sample evaporation during measurement can be significantly reduced by using a solvent trap.

6.1.14.3 The Vane Rheometer

The vane method uses a vane spindle attached to a constant stress rheometer. The spindle has four or eight blades attached to a cylindrical shaft. The depth of the suspension and the diameter of the vessel should be at least twice as large as the length and diameter of the vane to minimize slip effects at the boundary with the walls of the vessel. One can apply a constant shear rate and the torque is measured. Alternatively, one can apply a constant torque to the vane spindle and the strain is observed for a period of time at successively high stresses; the movement of the spindle is observed. When the applied stress exceeds the yield stress, the vane spindle starts to rotate. The vane method has the advantage of minimum disturbance of the structure when introducing the vane compared with rotational cylinders.

6.1.14.4 Cone and Plate Rheometer

This consists of a flat plate and a low-angle (θ) cone ($10^\circ \leq \theta \leq 40^\circ$) that rotates against the flat plate (Figure 6.14a).

The shear rate does not vary from the axis of rotation because both the linear velocity and the gap between the cone and plate increases with increase in distance from the rotational axis. If θ is small ($\sin\theta \approx \theta$), the shear rate is uniform across the gap:

$$\dot{\gamma} = \frac{r\Omega}{r\theta} = \frac{\Omega}{\theta} \tag{46}$$

Experimentally, one measures the moment of the force M_T (the torque) on the cone at a given shear rate. This is the sum of the forces on each element dr wide multiplied by the distance from the center R (the sum can be replaced by an integral from 0 to R):

$$M_T = \int_0^R \sigma 2\pi r^2 \,\mathrm{d}r = \frac{2\pi R^3}{3}\sigma \tag{47}$$

Three problems are encountered with the cone and plate geometry: (i) alignment (the tip of the cone must be at the plate center); (ii) the cone apex may be damaged; (iii) the particles or droplets may be 'ground' at the small gap at the center.

The above problems can be reduced by using a truncated cone as shown in Figure 6.14b; $R_1 \approx 0.2R$. The cone is aligned such that the 'apex' (which is now cut)

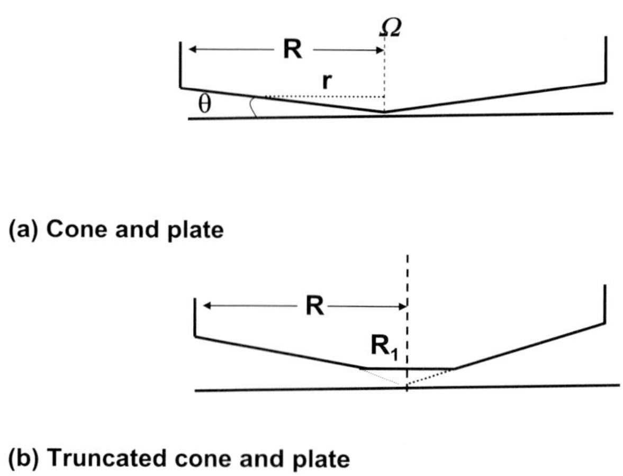

Figure 6.14 Schematic diagrams of the cone and plate system.

is still at the same position as the original cone without truncation. The torque is now given by

$$M_{\mathrm{T}} = \frac{2\pi(R^3 - R_1{}^3)}{3}\sigma \tag{48}$$

If $R_1 \approx 0.2R$, the torque is reduced by less than 1%.

6.1.14.5 Parallel Plates (Disks)

The parallel plate viscometer consists of two disks with an adjustable gap for holding the sample between the disks. Its advantage is the variable gap, which can be adjusted to accommodate dispersions with large particles, and calculation of the actual shear rate from the shear rates measured at different gap widths.

The torque generated on the upper disk is measured as a function of the angular velocity Ω of the lower disk. The maximum shear rate at the disk edge is given by

$$\dot{\gamma}_{\mathrm{m}} = \frac{\Omega R}{l} \tag{49}$$

where R is the radius of the disk and l is the gap width.

The torque can be obtained by using the same principle as for the cone and plate system, i.e. using equation 47 and substituting the stress by $\eta\gamma$ and using equation 49 for the shear rate, one obtains the following equation for a Newtonian:

$$M = \frac{2\pi\Omega\eta}{L}\int_0^R r^3\,\mathrm{d}r = \frac{\pi\Omega R^4\eta}{2L} \tag{50}$$

or a non-Newtonian one obtains an apparent viscosity η_{a} and the maximum shear rate given by equation 50. A correction has to be made to obtain the viscosity at each shear rate:

$$\eta(\dot{\gamma}_{\mathrm{m}}) = \eta_{\mathrm{a}}\left(1 + \frac{m}{4}\right) \tag{51}$$

where

$$m = \frac{\mathrm{d}\log\eta_{\mathrm{a}}}{\mathrm{d}\log\dot{\gamma}_{\mathrm{m}}} \tag{52}$$

6.1.14.6 The Brookfield Viscometer

This is the most commonly used viscometer due to the ease with which a viscosity can be measured (particularly for quality control). Several spindles are available to cover a wide range of viscosities and the results are usually quoted as viscosity at a given rpm using a specific spindle. For calculation of the shear rate, one has to use the long-cylinder spindle (as shown in Figure 6.15). This geometry approximates that of an infinitely long cylinder in an infinite sea of fluid.

For a 'long' cylinder, $L/R > 10$, where L is the length of the cylinder and R is its radius. For an 'infinite' medium, R_1/R_2 is very small. Let us denote R_1/R_2 by s.

In the instrument, one measures the apparent viscosity η_a as a function of rotation in rpm, denoted v:

$$v = \frac{30}{\pi}\Omega \tag{53}$$

The stress σ_1 at the cylinder is given by

$$\sigma_1 = \frac{M}{2\pi {R_1}^2 L} \tag{54}$$

The apparent viscosity η_a is given by

$$\eta_a = \frac{1}{4\pi {R_1}^2}\frac{M}{\Omega} = \frac{\sigma_1}{2\Omega} \tag{55}$$

Ω is given by equation 53 and σ_1 is then given by

$$\sigma_1 = \frac{\pi v \eta_a}{15} \tag{56}$$

Let

$$N = \frac{\mathrm{d}\log\eta_a}{\mathrm{d}\log\sigma_1} = \frac{1}{1+m}$$

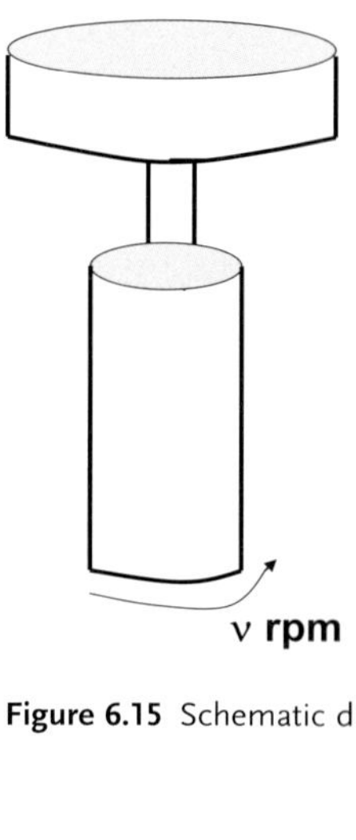

Figure 6.15 Schematic diagram of the Brookfield viscometer (cylinder).

where

$$m = \frac{\mathrm{d}\log \eta_\mathrm{a}}{\mathrm{d}\log \nu}$$

Then

$$\dot{\gamma}(\sigma_1) = \frac{2N\Omega}{1 - s^{-2N}} = 2N\Omega = \frac{\pi \nu}{15(1+m)} \tag{57}$$

$$\eta(\sigma_1) = \frac{\sigma_1}{\dot{\gamma}(\sigma_1)} = \eta_\mathrm{a}(1+m) \tag{58}$$

A plot of log η_a versus log ν is shown schematically in Figure 6.16 and one can obtain the shear rate and viscosity at each point using equations 57 and 58. Details of the analysis were given by Pierce [9].

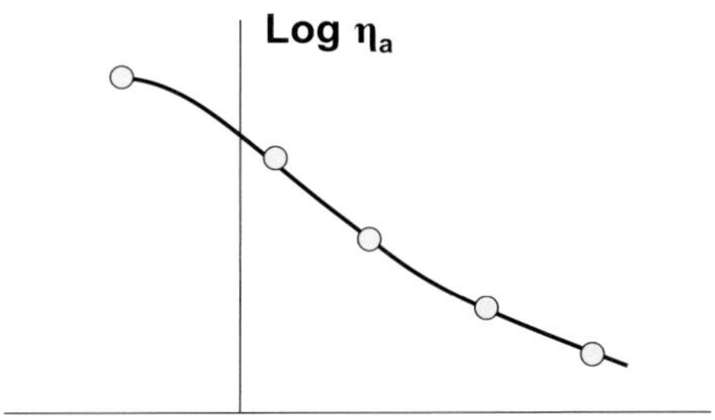

Figure 6.16 Log η_a versus log ν for a Brookfield viscometer.

6.2 Principles of Viscoelastic Behavior

6.2.1 Introduction

Elasticity deals with the mechanical properties of elastic solids which obey Hooke's law: stress $\sigma \propto$ strain γ (in small deformations), i.e.

$$\sigma = G'\gamma \tag{59}$$

where G' = shear modulus (Pa) and is independent of the applied strain (at low values).

Viscosity deals with the properties of liquids in the classical theory of hydrodynamics according to Newton's law: stress $\sigma \propto$ shear rate (at small shear rates), i.e.

$$\sigma = \eta\dot{\gamma} \tag{60}$$

where η = viscosity (Pa s) and is independent of the applied shear rate (at low values).

6.2.1.1 The Deborah Number

Viscous flow is a manifestation of the decay of elastically stored energy (Maxwell concept). Consider a dispersion in which all particles or droplets have time to diffuse to some low-energy state. If we strain (deform) the system, the structure is perturbed and the particles or droplets are now in a higher energy state. If the system is held in this new shape, the particles or droplets will diffuse until the original low-energy state is reached although the original shape is lost, i.e. viscous flow has occurred. The characteristic time that it takes for this process to occur is simply the stress relaxation time τ of the dispersion, which can be related to the experimental time t by the dimensionless Deborah number, D_e:

$$D_e = \frac{\tau}{t} \tag{61}$$

Since most rheological time-scale experiments are in the range 10^{-3} (1 ms)–10^3 (1 ks), then one can make a rough distinction between elastic (with long relaxation times), viscous (with very short relaxation times) and viscoelastic (with relaxation times comparable to the experimental time of measurement):

$D_e \gg 1$	$D_e \approx 1$	$D_e \ll 1$
elastic	viscoelastic	viscous

Viscoelasticity can be investigated using three different types of experiments [10–13]:

1. strain relaxation after sudden application of stress (creep)
2. stress relaxation after sudden application of strain
3. dynamic (oscillatory) technique.

The basic principles involved in each of the above methods are described in detail below, followed by the experimental methods that can be applied.

6.2.2 Strain Relaxation After Sudden Application of Stress (Creep)

A constant stress σ is applied on the system (which may be placed in the gap between two concentric cylinders or a cone and plate geometry) and the strain (relative deformation) γ or compliance J $(= \gamma/\sigma$, Pa$^{-1})$ is followed as a function of time for a period t. At $t = t$, the stress is removed and the strain γ or compliance J is followed for another period t.

The above procedure is referred to as 'creep measurement'. From the variation of J with t when the stress is applied and the change of J with t when the stress is removed (in this case J changes sign), one can distinguish between viscous, elastic and viscoelastic response, as illustrated in Figure 6.17.

- *Viscous response:* In this case, the compliance J shows a linear increase with increase in time, reaching a certain value after time t. When the stress is removed after time t, J remains the same, i.e. in this case no creep recovery occurs.
- *Elastic response:* In this case, the compliance J shows a small increase at $t = 0$ and it remains almost constant for the whole period t. When the stress is removed, J changes sign and it reaches zero after some time t, i.e. complete creep recovery occurs in this case.
- *Viscoelastic response:* At $t = 0$, J shows a sudden increase, followed by a slower increase for the time applied. When the stress is removed, J changes sign and shows an exponential decrease with increase in time (creep recovery), but it does not reach zero as in the case of an elastic response.

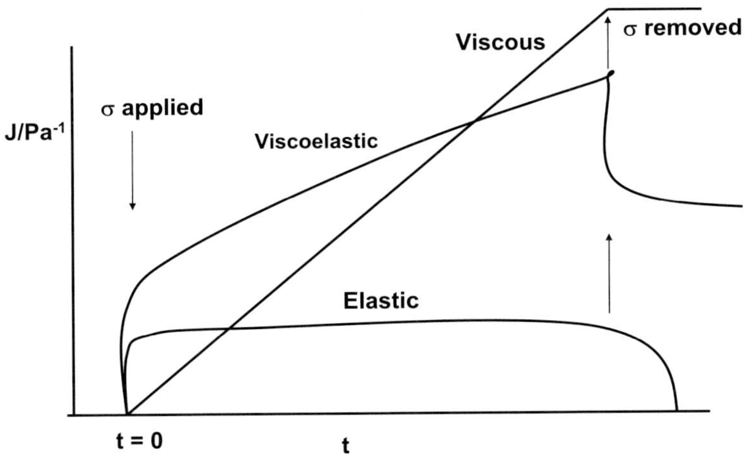

Figure 6.17 Creep curves for viscous, elastic and viscoelastic response.

6.2.3 Analysis of Creep Curves

6.2.3.1 Viscous Fluid

The linear curve of J versus t gives a slope that is equal to the reciprocal of viscosity:

$$J(t) = \frac{\gamma}{\sigma} = \frac{\dot{\gamma}t}{\sigma} = \frac{t}{\eta(0)} \tag{62}$$

This system can be represented by a mechanical analogue of a 'dash-pot' (Newtonian behavior), as illustrated schematically in Figure 6.18.

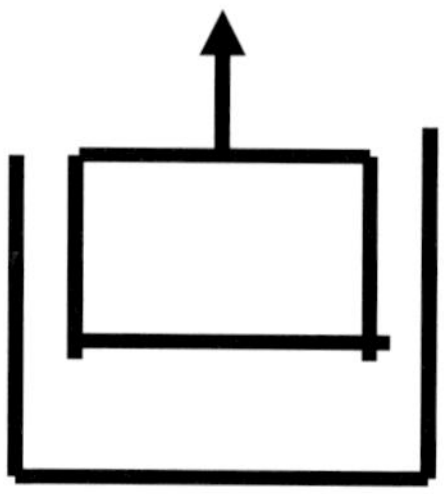

Figure 6.18 Schematic representation of a 'dash-pot'.

6.2.3.2 Elastic Solid

The increase in compliance at $t = 0$ (rapid elastic response), $J(t)$, is equal to the reciprocal of the instantaneous modulus $G(0)$:

$$J(t) = \frac{1}{G(0)} \tag{63}$$

The system can be represented by a mechanical analogue of a spring with a constant $G(0)$, as represented schematically in Figure 6.19.

Figure 6.19 Schematic representation of an elastic element.

6.2.4 Viscoelastic Response

6.2.4.1 Viscoelastic Liquid

Figure 6.20 shows the case for a viscoelastic liquid where the compliance $J(t)$ is given by two components: an elastic component J_e that is given by the reciprocal of the instantaneous modulus and a viscous component J_v that is given by $t/\eta(0)$:

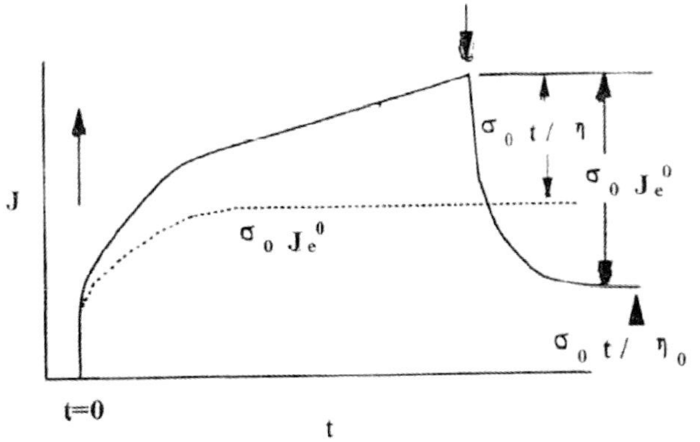

Figure 6.20 Creep curve for a viscoelastic liquid. Creep is the sum of a constant value $J_e\sigma_0$ (elastic part) and a viscous contribution $\sigma_0 t/\eta_0$.

$$J(t) = \frac{1}{G(0)} + \frac{t}{\eta(0)} \tag{64}$$

Figure 6.20 also shows the recovery curve, which gives $\sigma_0 {J_e}^0$, and when this is subtracted from the total compliance it gives $\sigma_0 t/\eta(0)$.

The mechanical analogue for a viscoelastic liquid is a spring and dash-pot in series (Maxwell element), as illustrated in Figure 6.21. The driving force for relaxation is spring and the viscosity controls the rate.

The Maxwell relaxation time τ_M is given by

$$\tau_M = \frac{\eta(0)}{G(0)} \tag{65}$$

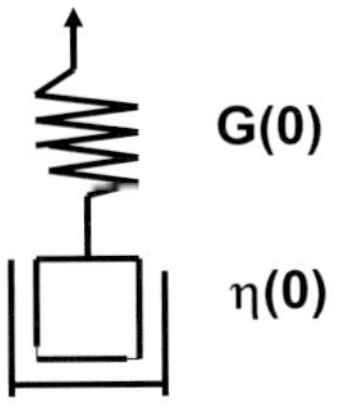

Figure 6.21 Mechanical analogue of a viscoelastic liquid.

6.2.4.2 **Viscoelastic Solid**

Here, complete recovery occurs as illustrated in Figure 6.22. The mechanical analogue in this case is represented by a spring and dash-pot in parallel as is illustrated in Figure 6.23.

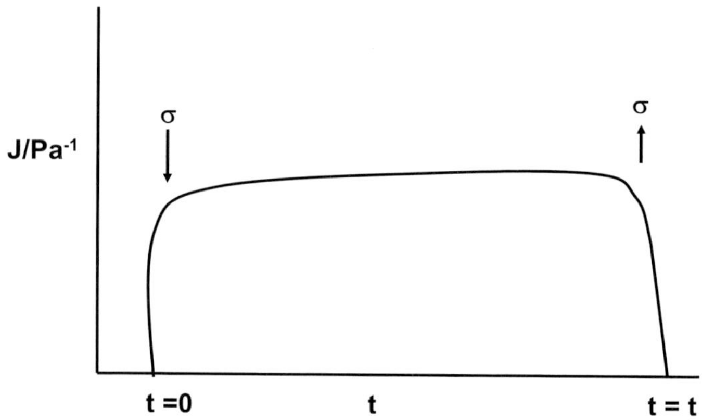

Figure 6.22 Creep curve for a viscoelastic solid.

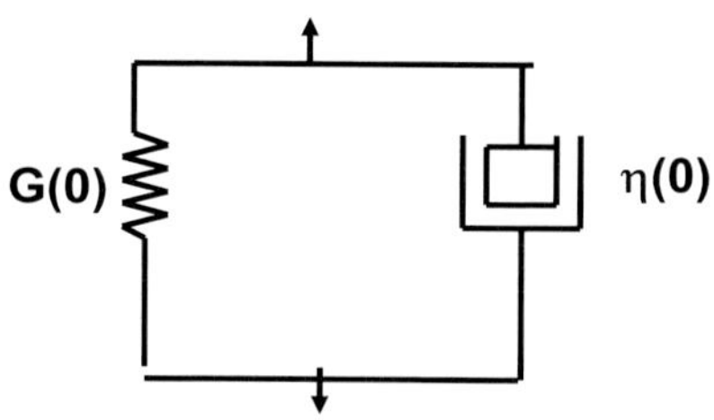

Figure 6.23 Mechanical analogue of a viscoelastic solid.

The dash-pot simply retards the motion of the spring (cf. suspension unit in a motor car) and the system is characterized by a Kelvin retardation time τ_k that is also given by the ratio $\eta(0)/G(0)$.

6.2.5
The Berger Model (Maxwell + Kelvin)

This model, which represents most practical dispersions, consists of a Maxwell element and a Kelvin element. The modulus of the spring in the Maxwell element is G_1 and the viscosity in the dash-pot is η_1. The Maxwell relaxation time is η_1/G_1. The modulus of the spring in the Kelvin element is G_2 and the viscosity in the dash-pot is η_2. The Kelvin retardation time is η_2/G_2. The Berger model gives an instantaneous elastic response from G_1 and a continuous viscous response from η_1.

More complex models can be introduced. In the generalized Maxwell model, several elements with different relaxation times are introduced (Figure 6.24). The generalized Kelvin model also consists of several Kelvin elements with different retardation times (Figure 6.25).

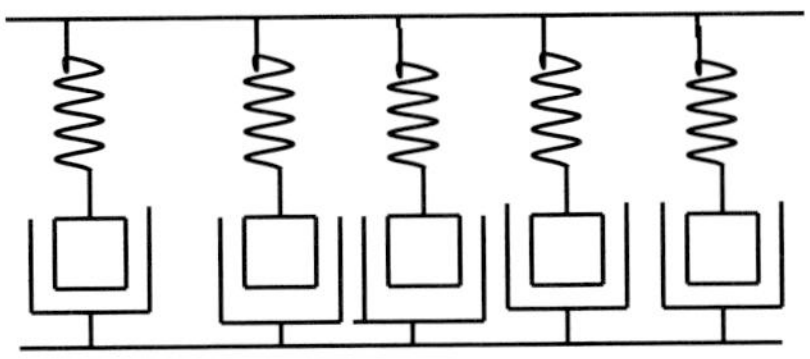

Figure 6.24 Generalized Maxwell model.

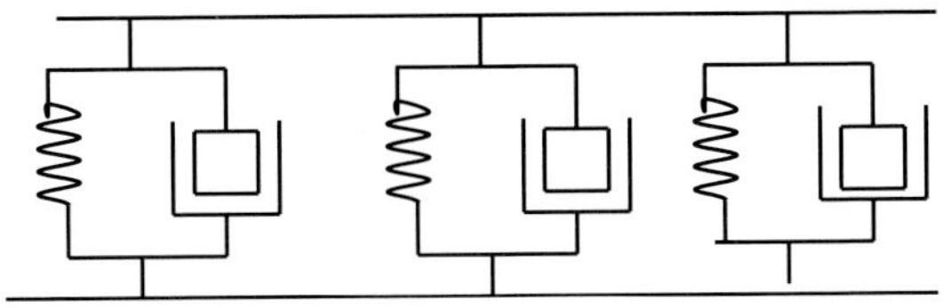

Figure 6.25 Generalized Kelvin model.

6.2.6 Creep Procedure

In creep experiments, one starts with a low applied stress (below the critical stress σ_{cr}; see below) at which the system behaves as a viscoelastic solid with complete recovery, as illustrated in Figure 6.22. The stress is gradually increased and several creep curves are obtained. Above σ_{cr}, the system behaves as a viscoelastic liquid showing only partial recovery, as illustrated in Figure 6.20. Figure 6.26 shows a schematic representation of the variation of compliance J with time t at increasing σ (above σ_{cr}).

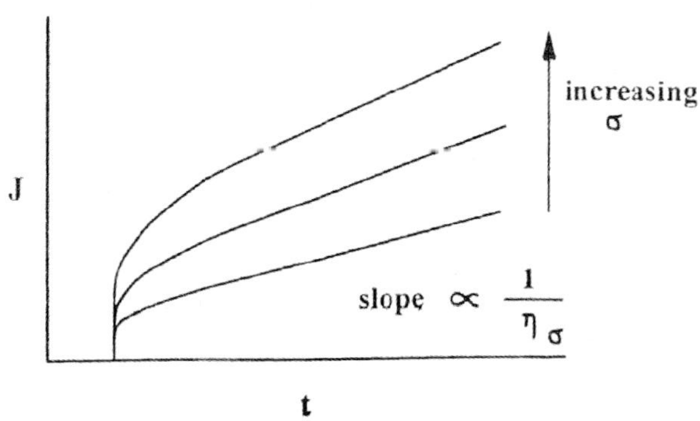

Figure 6.26 Creep curves at increasing applied stress. Creep measurements (constant stress) can be used to obtain the residual or zero shear viscosity.

From the slopes of the lines, one can obtain the viscosity η_σ at each applied stress. A plot of η_σ versus σ is shown in Figure 6.27. This shows a limiting viscosity $\eta(0)$ below σ_{cr} and above σ_{cr} the viscosity shows a sharp decrease with further increase in σ.

$\eta(0)$ is referred to as the residual or zero shear viscosity, which is an important parameter for predicting sedimentation. σ_{cr} is the critical stress above which the structure 'breaks down'. It is sometimes referred to as the 'true' yield stress.

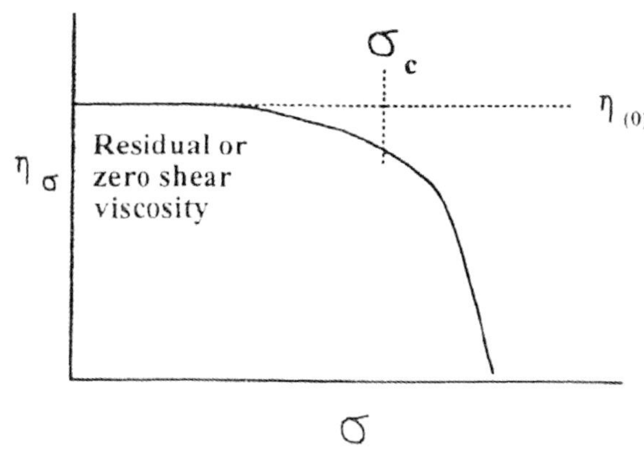

Figure 6.27 Variation of viscosity with applied stress. Critical stress is a useful parameter (related to yield stress) as it denotes the stress at which the structure 'breaks down'.

6.2.7
Stress Relaxation After Sudden Application of Strain

In this case, a small strain is rapidly applied within a very short period of time (which must be smaller than the relaxation time of the system) and is kept at a constant value. The shear rate remains constant within this period. This is illustrated in Figure 6.28.

The stress will follow the strain and increases to a maximum value $\sigma(0)$. For a perfectly elastic material, $\sigma(0)$ remains constant over time t. For a viscoelastic liquid, the stress decreases exponentially with time, reaching zero at infinite time. The stress required to maintain a constant strain decreases with time due to viscous flow. This is illustrated in Figure 6.29. For a viscoelastic solid, the stress reaches a limiting value at infinite time. The variation of stress with time is similar to a kinetic process represented by first-order equations.

The stress $\sigma(t)$ is related to the initial maximum stress $\sigma(0)$ by

$$\sigma(t) = \sigma(0) \exp\left(-\frac{t}{\tau_m}\right) \tag{66}$$

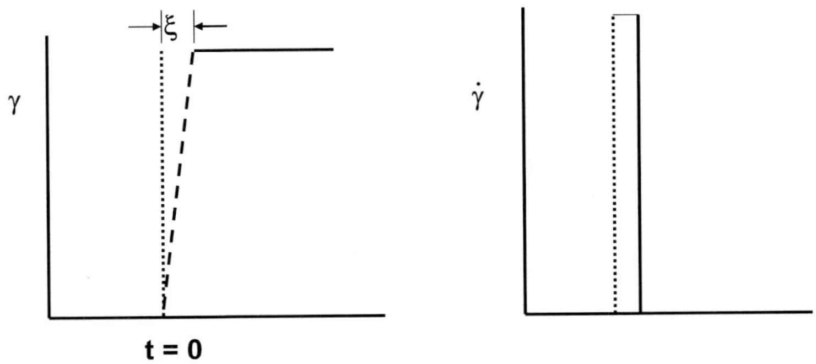

Figure 6.28 Schematic representation of a strain experiment.

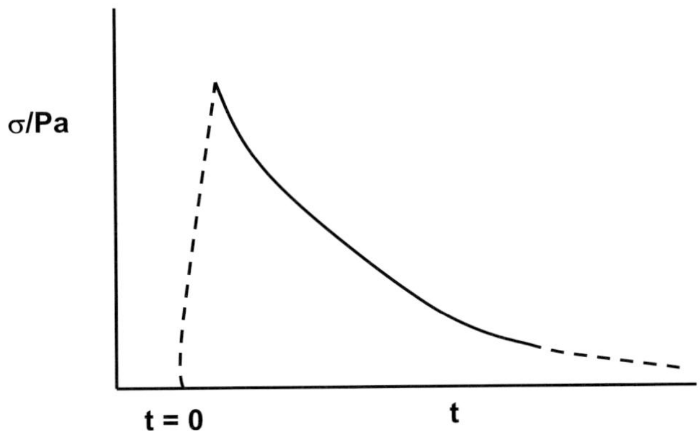

Figure 6.29 Stress relaxation after sudden application of strain.

where τ_{m} is the Maxwell relaxation time, which is given by the ratio of the viscosity η to the modulus G:

$$\tau_{\mathrm{m}} = \frac{\eta}{G} \tag{67}$$

If the shear stress in equation 66 is divided by the applied strain γ, one obtains the shear modulus $G(t)$:

$$G(t) = \frac{\sigma(t)}{\gamma} = \frac{\sigma(0)}{\gamma} \exp\left(-\frac{t}{\tau_{\mathrm{m}}}\right) = G(0) \exp\left(-\frac{t}{\tau_{\mathrm{m}}}\right) \tag{68}$$

Figure 6.30 shows the variation of the modulus G with time for a viscoelastic liquid and Figure 6.31 shows the trend for a viscoelastic solid.

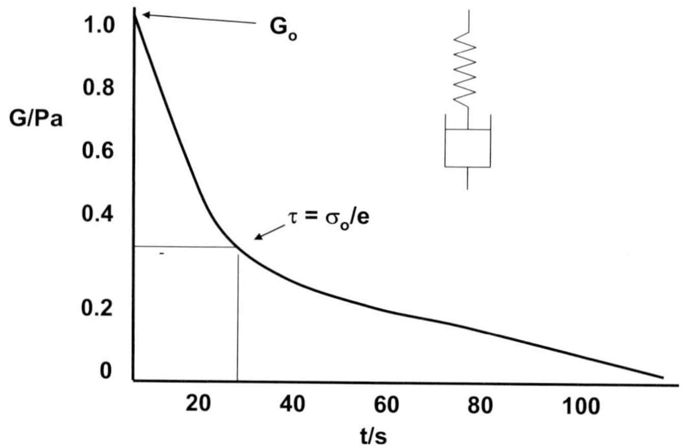

Figure 6.30 Variation of modulus G with time for a viscoelastic liquid.

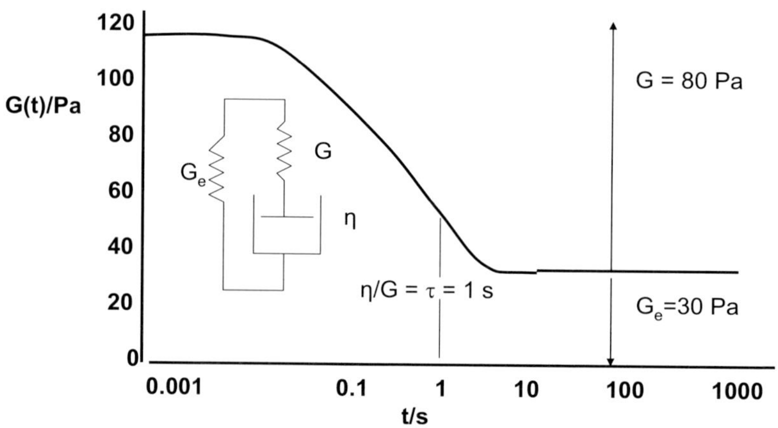

Figure 6.31 Variation of $G(t)$ with time for a viscoelastic solid.

For a viscoelastic solid, the modulus reaches a limiting value G_e at long time (sometimes referred to as the equilibrium modulus). In this case, equation 68 has to be modified to account for G_e:

$$G(t) = G(0)\exp\left(-\frac{t}{\tau_m}\right) + G_e \tag{69}$$

Note that according to equations 66 and 68 $t = \tau_m$ when $\sigma(t) = \sigma(0)/e$ or when $G(t) = G(0)/e$. This shows that stress relaxation can be used to obtain the relaxation time for a viscoelastic liquid.

6.2.8
Dynamic (Oscillatory) Techniques

These concern the response of the material to an oscillating stress or strain. When a sample is constrained in, say, a cone and plate or concentric cylinder assembly, an oscillating strain at a given frequency ω (rad s^{-1}) ($\omega = 2\nu\pi$, where ν is the frequency in cycles per second or Hz) can be applied to the sample. After an initial start-up period, a stress develops in response to the applied strain, i.e. it oscillates with the same frequency. The change of the sine waves of the stress and strain with time can be analyzed to distinguish between elastic, viscous and viscoelastic response. Analysis of the resulting sine waves can be used to obtain the various viscoelastic parameters, as discussed below.

Three cases can be considered:

- *Elastic response*, whereby the maximum of the stress amplitude is at the same position as the maximum of the strain amplitude (no energy dissipation). In this case, there no time shift between stress and strain sine waves.
- *Viscous response*, whereby the maximum of the stress is at the point of maximum shear rate (i.e. the inflection point) where there is maximum energy dissipation. In this case, the strain and stress sine waves are shifted by $\omega t = \pi/2$ (referred to as the phase angle shift δ, which in this case is 90°).
- *Viscoelastic response*, whereby the phase angle shift δ is greater than 0° but less than 90°.

A schematic representation of the above three cases is shown in Figure 6.32.

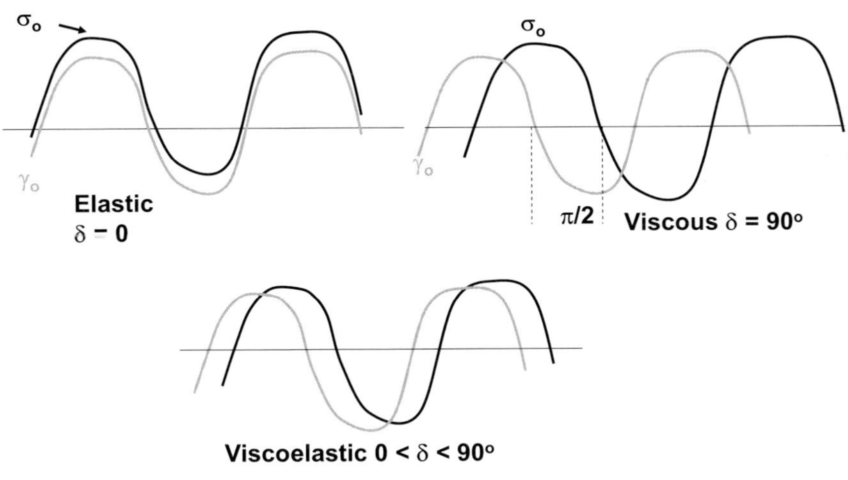

Figure 6.32 Schematic representation of the response to an oscillatory strain or stress for elastic, viscous and viscoelastic systems.

6.2.8.1 Analysis of Oscillatory Response for a Viscoelastic System

Let us consider the case of a viscoelastic system. The sine waves of strain and stress are shown in Figure 6.33. The frequency ω is in rad s^{-1} and the time shift between strain and stress sine waves is Δt. The phase angle shift δ is given by (in dimensionless units of radians)

$$\delta = \omega \Delta t \tag{70}$$

As discussed earlier, for a perfectly elastic solid $\delta = 0°$, for a perfectly viscous liquid $\delta = 90°$ and for a viscoelastic system $0° < \delta < 90°$.

The ratio of the maximum stress σ_0 to the maximum strain γ_0 gives the complex modulus $|G^*|$:

$$|G^*| = \frac{\sigma_0}{\gamma_0} \tag{71}$$

$|G^*|$ can be resolved into two components: the storage (elastic) modulus G', the real component of the complex modulus; and the loss (viscous) modulus G'', the imaginary component of the complex modulus:

$$|G^*| = G' + iG'' \tag{72}$$

where i is the imaginary number, equal to $(-1)^{\frac{1}{2}}$.

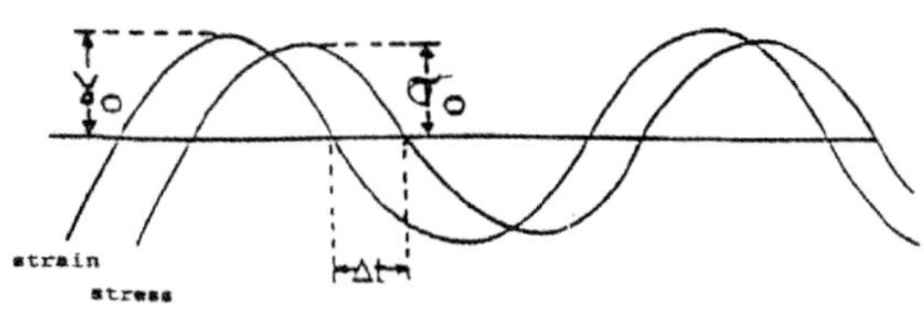

Δt = time shift for sine waves of stress and strain.

$\Delta t \omega = \delta$ phase angle shift

ω = frequency in radian s^{-1}

$\omega = 2\pi\nu$

Perfectly elastic solid	$\delta = 0$
Perfectly viscos liquid	$\delta = 90°$
Viscoelastic system	$0 < \delta < 90°$

Figure 6.33 Strain and stress sine waves for a viscoelastic system.

The complex modulus can be resolved into G' and G'' using vector analysis and the phase angle shift δ as shown below and in Figure 6.34.

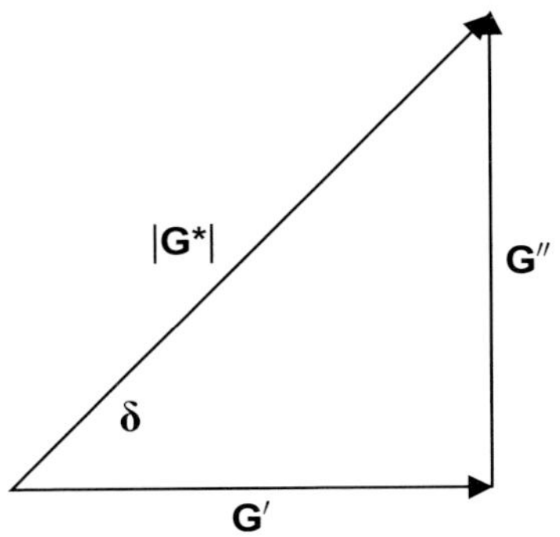

Figure 6.34 Resolution of the complex modulus.

6.2.8.2 Vector Analysis of the Complex Modulus

$$G' = |G^*| \cos \delta \tag{73}$$

$$G'' = |G^*| \sin \delta \tag{74}$$

$$\tan \delta = \frac{G''}{G'} \tag{75}$$

Dynamic viscosity η':

$$\eta' = \frac{G''}{\omega} \tag{76}$$

Note that $\eta \rightarrow \eta(0)$ as $\omega \rightarrow 0$.

Both G' and G'' can be expressed in terms of frequency ω and Maxwell relaxation time τ_m by

$$G'(\omega) = G\frac{(\omega\tau_m)^2}{1 + (\omega\tau_m)^2} \tag{77}$$

and

$$G''(\omega) = G\frac{\omega\tau_m}{1 + (\omega\tau_m)^2} \tag{78}$$

In oscillatory techniques, one has to carry out two types of experiments: (i) strain sweep, where the frequency ω is kept constant and G^*, G' and G'' are measured as a function of strain amplitude; and (ii) frequency sweep, where the strain is kept constant (in the linear viscoelastic region) and G^*, G' and G'' are measured as a function of frequency.

6.2.9
Strain Sweep

The frequency is fixed say at 1 Hz (or 6.28 rad s^{-1}) and G^*, G' and G'' are measured as a function of strain amplitude γ_0. This is illustrated in Figure 6.35. G^*, G' and G'' remain constant up to a critical strain γ_{cr}. This is the linear viscoelastic region where the moduli are independent of the applied strain. Above γ_{cr}, G^* and G' start to decrease whereas G'' starts to increase with further increase in γ_0. This is the non-linear region.

γ_{cr} may be identified as the critical strain above which the structure starts to 'break down'. It can also be shown that above another critical strain, G' becomes higher than G'. This is sometimes referred to as the 'melting strain', at which the system becomes more viscous than elastic.

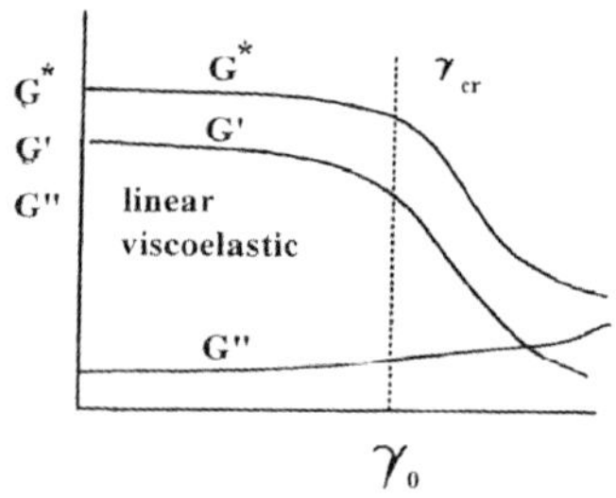

Figure 6.35 Schematic representation of strain sweep. Fixed frequency (0.1 or 1 Hz), and follow G^*, G' and G'' with strain amplitude γ_0. Linear viscoelastic region: G^*, G' and G'' are independent of strain amplitude.

6.2.10
Oscillatory Sweep

The strain γ_0 is fixed in the linear region (taking a mid-point, i.e. not a too low strain where the results may show some 'noise' and far from γ_{cr}). G^*, G' and G'' are then measured as a function of frequency (a range of 10^{-3}–10^2 rad s^{-1} may be chosen depending on the instrument and operator patience). Figure 6.36 shows a schematic representation of the variation of G^*, G' and G'' with frequency ω (rad s^{-1}) for a viscoelastic system that can be represented by a Maxwell model). One can identify a characteristic frequency ω^* at which $G' = G''$ (the 'cross-over point') which can be used to obtain the Maxwell relaxation time τ_m:

$$\tau_m = \frac{1}{\omega^*} \tag{79}$$

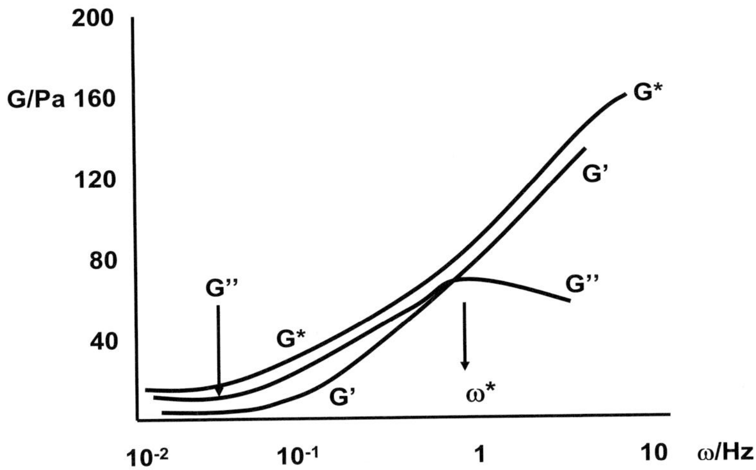

Figure 6.36 Schematic representation of oscillatory measurements for a viscoelastic liquid.

In the low-frequency regime, $\omega < \omega^*$, $G'' > G'$. This corresponds to a long-time experiment (time is the reciprocal of frequency) and hence the system can dissipate energy as viscous flow.

In the high-frequency regime, $\omega > \omega^*$, $G' > G''$. This corresponds to a short-time experiment where energy dissipation is reduced.

At sufficiently high frequency $G' \gg G''$, $G'' \rightarrow 0$ and $G' \approx G^*$. The high-frequency modulus $G'(\infty)$ is sometimes referred to as the 'rigidity modulus' where the response is mainly elastic.

For a viscoelastic solid, G' does not become zero at low frequency. G'' still shows a maximum at intermediate frequency. This is illustrated in Figure 6.37.

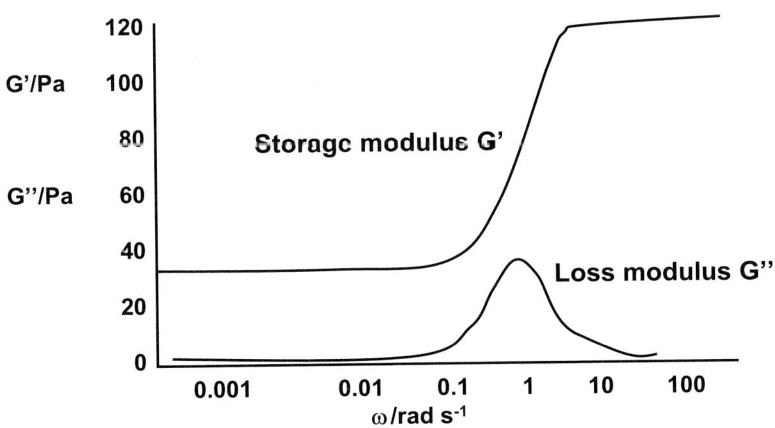

Figure 6.37 Schematic representation of oscillatory measurements for a viscoelastic solid.

6.2.11
The Cohesive Energy Density, E_c

The cohesive energy density, which is an important parameter for identification of the 'strength' of the structure in a dispersion, can be obtained from the change of G' with γ_0 (see Figure 6.35):

$$E_c = \int_0^{\gamma_{cr}} \sigma \, d\gamma \tag{80}$$

where σ is the stress in the sample, which is given by

$$\sigma = G'\gamma \tag{81}$$

$$E_c = \int_0^{\gamma_{cr}} G'\gamma_{cr} \, d\gamma = \frac{1}{2}\gamma_{cr}{}^2 G' \tag{82}$$

where E_c is in J m^{-3}.

6.2.12
Viscoelastic Measurements

Viscoelastic measurements (low deformation) require the use of sensitive rheometers that operate using an air bearing (drag-cup motor) in order to minimize friction (very low torques need to be applied).

Three types of measurements are performed (which are equivalent but not identical) and these are complementary in their application:

- constant stress (creep) measurements
- constant strain (stress relaxation) measurements
- dynamic (oscillatory) measurements.

Several commercial rheometers are available to perform the above measurements, e.g. from Bohlin Instruments (Malvern, UK), CarriMed and Rheometric Instruments (TA Instruments, UK), Physica (Paar Physica, Germany) and Haake (Germany).

6.2.12.1 Constant Stress (Creep) Measurements

In these measurements, a constant stress σ is applied on the sample (usually electrically applied via a motor attached to the drag-cup) and the strain $\gamma(t)$ or compliance $J(t)$ [$= \gamma(t)/\sigma$ in Pa^{-1}] is measured using a position sensor below the drag-cup motor). This is illustrated schematically in Figure 6.38, which shows both controlled stress and controlled strain set-ups.

A controllable, known torque is applied, say, to the cone and the resulting rate of rotation of the plate is measured. Temperature control of the sample is usually achieved using a Peltier device. A solvent trap for the cone and plate (which is most commonly used) is placed to cover the whole geometry.

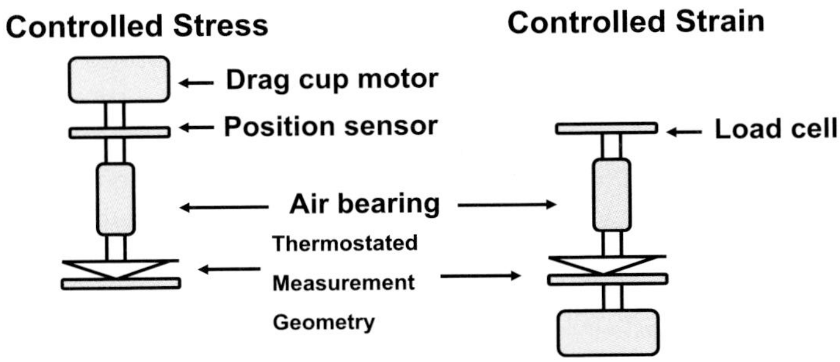

Figure 6.38 Controlled stress and controlled strain set-ups.

The main requirements for constant stress measurements are as follows:

1. The applied torque should cover a reasonable range to allow one to increase the stress until flow occurs.
2. Low bearing fraction (the frictional torque should be 2–3 orders of magnitude lower than the applied torque).
3. The instrument should be fully computer controlled (with easy software).
4. The output of the $J(t)$ versus t curves should be automatically recorded until a steady state is reached [linear increase of $J(t)$ with t].
5. Software should automatically calculate the viscosity η_σ at every σ applied and plots of η_σ versus σ should be automatically displayed to obtain the limiting (residual or zero shear) viscosity $\eta(0)$ and critical stress ('true' yield value) σ_{cr}.

The procedure for constant stress measurements is as follows. For a given sample, one would not know *a priori* the critical stress σ_{cr} below which the material shows a near 'elastic' response (whereby the compliance shows a very slow

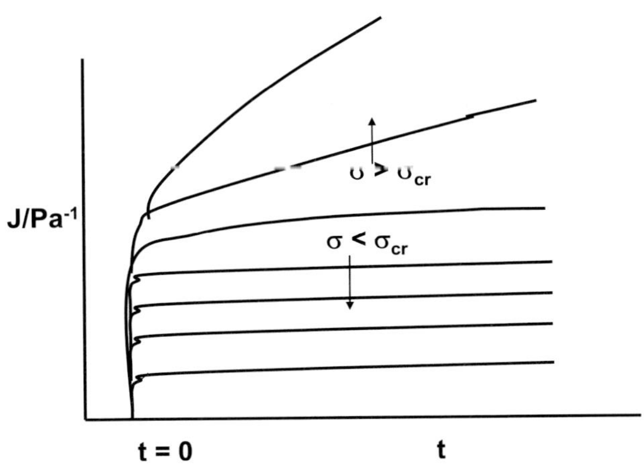

Figure 6.39 Schematic representation of creep curves.

increase with time after the instantaneous increase and complete recovery after release of the stress). The slopes of $J(t)$ versus t below σ_{cr} are very small and equal and this gives $\eta(0)$, which is the reciprocal of this small slope (high viscosity values are obtained below σ_{cr}).

Initially, one should use a sample to obtain σ_{cr} roughly by applying large increments of σ (a logarithmic scale can be used). Once σ_{cr} has been roughly obtained, measurements are then carried on a new sample where reasonable σ values are applied below and above σ_{cr}. In this way, several creep curves can be measured (about 10 curves are usually sufficient at σ values below and above σ_{cr}. A schematic representation of the creep curves is shown in Figure 6.39.

6.2.12.2 **Stress Relaxation (After Application of Constant Strain)**

In this case, the material is kept under constant deformation (constant shear rate) and the resulting stress is measured as a function of time. The main problem is the speed with which a sudden strain is applied. This should be done in a much shorter time than the relaxation time of the sample. Very few strain-controlled instruments are available commercially, of which the 'Fluid' rheometer of Rheometrics is perhaps the only one available instrument (Bohlin used to have a strain-controlled instrument, namely the Bohlin VOR, but this is no longer available).

One of the earliest strain-controlled instruments was the Weissenberg rheogoniometer with a cone and plate attachment. This is now commercially available from TA Instruments.

One of the main problems with stress relaxation experiments is the possibility of fluctuations in the stress at short times, in particular for elastic samples. The above behavior can be explored by using 'stress growth' experiments. The application of a linearly ramped strain can provide information on both the sample viscosity and elasticity. The stress will grow in proportion to the applied strain. The ratio of the strain to the applied time gives the shear rate. Figure 6.40 shows the

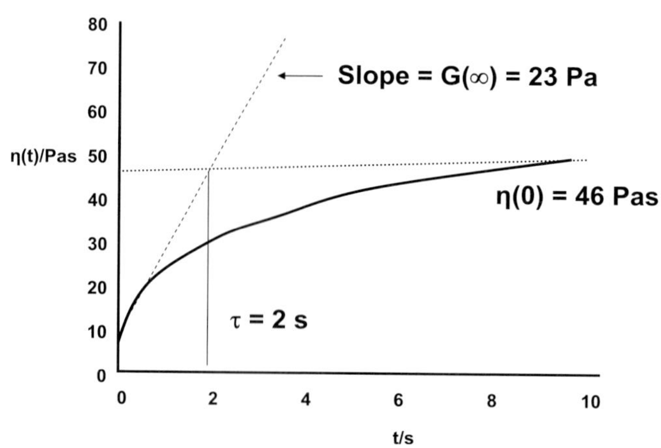

Figure 6.40 Stress growth function for a Maxwell model.

variation of viscosity $\eta(t)$ versus time for a system represented by a Maxwell model with a relaxation time τ_r. The results can be analyzed to obtain the modulus $G(\infty)$ and the zero shear viscosity $\eta(0)$.

6.2.12.3 Dynamic (Oscillatory) Measurements

These refer to experiments in which both stress and strain vary harmonically. Suppose that a harmonic unique shear stress of amplitude σ_0 is applied to the upper face of a thin block of material. The stress σ will produce a strain γ_0 that is shifted by an angle δ (the phase angle shift) according to

$$\sigma = \sigma_0 \cos \omega t \tag{83}$$

$$\gamma = \gamma_0(\cos \omega t - \delta) \tag{84}$$

The above stress and strain wave forms are shown schematically in Figure 6.41. From σ_0, γ_0 and δ, one can calculate the storage modulus G' (the elastic component) and the loss modulus G'' (the viscous component):

$$G' = \frac{\sigma_0}{\gamma_0} \cos \delta \tag{85}$$

$$G'' = \frac{\sigma_0}{\gamma_0} \sin \delta \tag{86}$$

In dynamic measurements, the stress is proportional to the force F applied to the test material. The strain is proportional to the relative displacement of the bounding surfaces. The shear rate is proportional to the relative velocity v. If the force and displacement vary harmonically,

$$F = bG^* x = b\eta^* v = \frac{bG^*}{i\omega} v \tag{87}$$

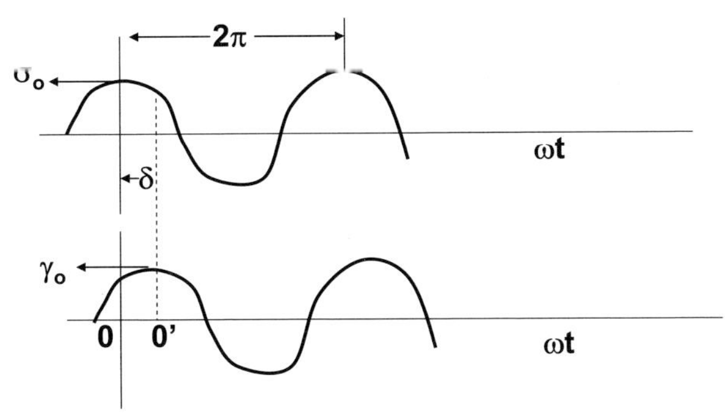

Figure 6.41 Stress and strain wave forms.

where F, x and v represent complex amplitudes and b is a shape factor for the test material. The material can be confined between cone and plate, parallel plates or cylinders. One oscillates one of the boundary surfaces and then measures the motion of the other surface.

In the Bohlin VOR, for example when using a concentric cylinder, the inner cylinder (the suspended member) is connected to interchangeable torque bars (to measure the stress). A sinusoidal strain is applied to the cup by moving it back and forth in a sinusoidal manner. The displacement is measured using a transducer.

If θ_2 and θ_1 are the angular positions of driven and suspended elements, respectively, then the torque in the material M is

$$M = bG^*(\theta_2 - \theta_1) \tag{88}$$

$$G' = \frac{1}{b}\frac{|M|}{\theta_2}\cos\phi \tag{89}$$

$$G'' = \frac{1}{b}\frac{|M|}{\theta_2}\sin\phi \tag{90}$$

where ϕ is the phase shift of the torque element relative to θ_2. Normally one measures the ratio θ_1/θ_2 and the phase shift ϕ of suspended element to obtain G' and G''. The computation is carried out by the software of the computer.

As mentioned earlier, two types of measurements are carried out:

1. *Strain sweep measurements:* the frequency is fixed at 1 Hz (or 6.28 rad s^{-1}) and the strain amplitude is gradually increased from the lowest possible value to a critical value γ_{cr} (which depends on the system) above which the system shows a non-linear response.
2. *Oscillatory sweep measurements:* the strain is fixed below γ_{cr} (i.e. in the linear viscoelastic region where the moduli are independent of the applied strain) and measurement is then carried out as a function of frequency. The range used depends on the system and the instrument. Mostly measurements are carried out in the range 10^{-2}–1 Hz.

6.2.12.4 **Shear Modulus (Rigidity) Measurement**

The shear (rigidity) modulus can be measured by initiating a shear wave in the dispersion and measuring the damping of the wave [13, 14]. A simple set-up is the Rank Brothers (Bottisham, Cambridge, UK) pulse shearometer. Two disks (Perspex or steel) are placed in the dispersion and the gap between the disks can be changed using a micrometer. Each disk is connected to a transducer (A and B): piezoelectric crystals of LiCl. B is connected to a pulse generator that provides a small-amplitude ($<10^{-4}$ rad) deformation with high frequency ($\sim$200 Hz). The output from A is displayed using a computer to observe the shear wave. This is shown schematically in Figure 6.42.

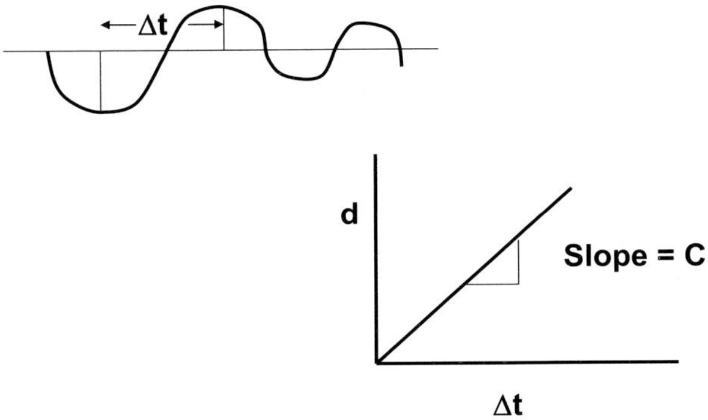

Figure 6.42 Schematic representation of a shear wave and variation of d with Δt.

A plot of d (the gap width between the plates) versus Δt is linear and the slope C gives the velocity of the shear wave. The rigidity (high-frequency) modulus $G(\infty)$ is simply given by the expression

$$G(\infty) = \rho C^2 \tag{91}$$

where ρ is the density of the suspension.

References

1 J.R. van Wazer, J.W. Lyons, K.Y. Kim and R.E. Cowell, *Viscosity and Flow Measurements*, Interscience, New York (1983).
2 R.W. Whorlow, *Rheological Techniques*, John Wiley and Sons, Inc., New York (1980).
3 H.A. Barnes, J.F. Hutton and K. Walters, *An Introduction to Rheology*, Elsevier, Amsterdam (1989)
4 J.W. Goodwin and R.W. Hughes, *Rheology for Chemists*, Royal Society of Chemistry, Cambridge (2000).
5 E.C. Bingham, *Fluidity and Plasticity*, McGraw Hill, New York (1922).
6 W.H. Herschel and R. Bulkley, *Proc. Am. Soc. Test Mater.*, *26*, 621 (1926); *Kolloid Z.*, *39*, 291 (1926).
7 N. Casson, in *Rheology of Disperse Systems*, ed. C.C. Mill, Pergamon Press, New York, pp. 84–104 (1959).
8 M.M. Cross, *J. Colloid Interface Sci.*, *20*, 417 (1965).
9 P.E. Pierce, *J. Paint Technol.*, *43*, 35 (1971).
10 J.D. Ferry, *Viscoelastic Properties of Polymers*, John Wiley & Sons, Inc., New York (1980).
11 J.W. Goodwin, in *Surfactants*, ed. Th.F. Tadros, Academic Press, London, (1984).
12 Th.F. Tadros, *Adv. Colloid Interface Sci.*, *68*, 97 (1996).
13 H. van Olphen, *Clays Clay Miner.*, *4*, 204 (1956).
14 J.W. Goodwin and A.M. Khider, in *Colloid and Interface Science*, ed. M. Kerker, Academic Press, New York, Vol. IV, p. 529 (1976).

7
Rheology of Suspensions, Emulsions and Their Mixtures (Suspoemulsions)

7.1
Rheology of Suspensions

7.1.1
Introduction

The rheology of suspensions depends on the balance between three main forces: Brownian diffusion; hydrodynamic interaction; and interparticle forces. The last aspect was discussed in detail in Chapter 4.

The above forces are determined by four main parameters: (i) the volume fraction ϕ (total volume of the particles divided by the volume of the dispersion), (ii, iii) the particle size and shape distribution and (iv) the net energy of interaction G_T, i.e. the balance between repulsive and attractive forces.

The earliest theory for the prediction of the relationship between the relative viscosity η_r and ϕ was described by Einstein [1] and is applicable to $\phi \leq 0.01$.

7.1.2
The Einstein Equation

Einstein [1] assumed that the particles behave as hard-spheres (with no net interaction). The flow field has to dilate because the liquid has to move around the flowing particles. At $\phi \leq 0.01$, the disturbance around one particle does not interact with the disturbance around another.

η_r is related to ϕ by the following expression:

$$\eta_r = 1 + [\eta]\phi = 1 + 2.5\phi \quad (1)$$

where $[\eta]$ is referred to as the intrinsic viscosity and has the value 2.5.

For the above hard-sphere very dilute dispersions, the flow is Newtonian, i.e. the viscosity is independent of shear rate. At higher ϕ values ($0.2 > \phi > 0.1$), one has to consider the hydrodynamic interaction suggested by Batchelor [2] that is still valid for hard spheres.

Colloids and Interface Science Series, Vol. 6
Colloids in Paints. Tharwat F. Tadros

ISBN: 978-3-527-31466-9

7.1.3
The Batchelor Equation

When $\phi > 0.01$, hydrodynamic interactions between the particles become important. When the particles come close to each other, the nearby stream lines and the disturbance of the fluid around one particle interacts with that around a moving particle.

Using the above picture, Batchelor [2] derived the following expression for the relative viscosity:

$$\eta_r = 1 + 2.5\phi + 6.2\phi^2 + O\phi^3 \tag{2}$$

The third term in equation 2, i.e. $6.2\phi^2$, is the hydrodynamic term, whereas the fourth term is due to higher order interactions.

7.1.4
Rheology of Concentrated Suspensions

When $\phi > 0.2$, η_r becomes a complex function of ϕ. At such high volume fractions, the system mostly shows non-Newtonian flow ranging from viscous to viscoelastic to elastic response, depending on the Deborah number, D_e:

$$D_e = \frac{\tau}{t_e} \tag{3}$$

where τ is the relaxation time of the system and t_e is the experimental time of measurement.

Three responses can be considered: (i) viscous response, $D_e < 1$; (ii) elastic response, $D_e > 1$; and (iii) viscoelastic response, $D_e \sim 1$. Clearly, these responses for any suspension depend on the time or frequency of the applied stress or strain (see Chapter 5).

Four different types of systems (with increasing complexity) can be considered, as described below:

1. *Hard-sphere suspensions:* These are systems where both repulsive and attractive forces are screened.
2. *Systems with 'soft' interaction:* These are systems containing electrical double layers with long-range repulsion. The rheology of the suspension is determined mainly by the double layer repulsion.
3. *Sterically stabilized suspensions:* The rheology is determined by the steric repulsion produced by adsorbed nonionic surfactant or polymer layers. The interaction can be 'hard' or 'soft' depending on the ratio of adsorbed layer thickness to particle radius (δ/R).
4. *Flocculated systems:* These are systems where the net interaction is attractive. One can distinguish between weak (reversible) and strong (irreversible) flocculation, depending on the magnitude of the attraction.

7.1.5
Rheology of Hard-Sphere Suspensions

Hard-sphere suspensions (neutral stability) were developed by Krieger and Dougherty [3, 4] using polystyrene latex suspensions whereby the double layer repulsion was screened by using NaCl or KCl at a concentration of 10^{-3} mol dm^{-3} or replacing water by a less polar medium such as benzyl alcohol.

The relative viscosity η_r $(= \eta/\eta_0)$ is plotted as a function of reduced shear rate $\dot{\gamma}_{red}$ (shear rate $\times$ time for a Brownian diffusion t_r):

$$\dot{\gamma}_{red} = \dot{\gamma} t_r = \frac{6\pi \dot{\gamma} a^3}{kT} \tag{4}$$

where a is the particle radius, η_0 is the viscosity of the medium, k is the Boltzmann constant and T is the absolute temperature.

A plot of η/η_0 versus $\eta_0 a^3/kT$ is shown in Figure 7.1 at $\phi = 0.4$ for particles with different sizes. At a constant ϕ all points fall on the same curve. The curves are shifted to higher values for larger ϕ and to lower values for smaller ϕ.

The curve in Figure 7.1 shows two limiting (Newtonian) viscosities at low and high shear rates that are separated by a shear thinning region. In the low shear rate regime, the Brownian diffusion predominates over hydrodynamic interaction and the system shows a 'disordered' three-dimensional structure with high relative viscosity. As the shear rate is increased, these disordered structures starts to form layers coincident with the plane of shear and this results in the shear thinning region. In the high shear rate regime, the layers can 'slide' freely and hence a Newtonian region (with much lower viscosity) is obtained. In this region, the hydrodynamic interaction predominates over the Brownian diffusion.

If the relative viscosity in the first or second Newtonian region is plotted versus the volume fraction, one obtains the curve shown in Figure 7.2.

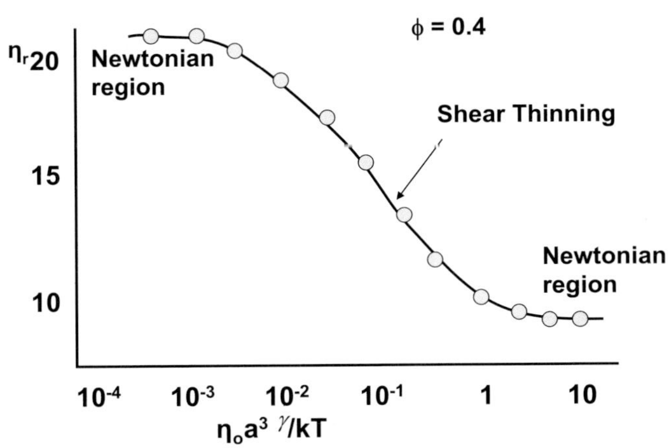

Figure 7.1 Reduced viscosity versus reduced shear rate for hard-sphere suspensions.

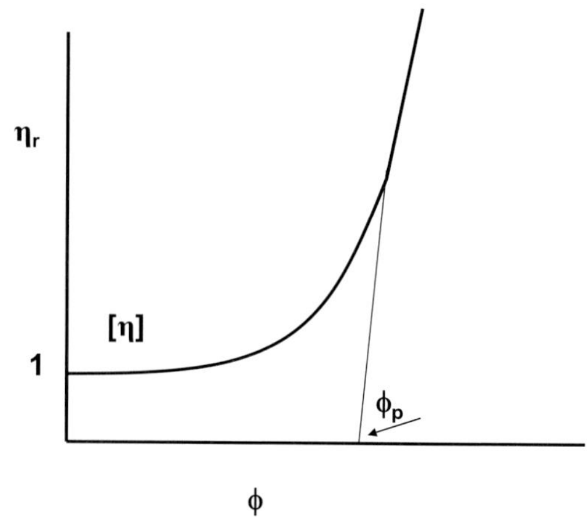

Figure 7.2 Relative viscosity versus volume fraction for hard-sphere suspensions.

The curve in Figure 7.2 has two asymptotes. The slope of the linear portion at low ϕ values (the Einstein region) gives the intrinsic viscosity $[\eta]$, which is equal to 2.5. The other asymptote occurs at a critical volume fraction ϕ_p at which the viscosity shows a sharp increase with increase in ϕ. ϕ_p is referred to as the maximum packing fraction for hard spheres. For hexagonal packing of equal-sized spheres $\phi_p = 0.74$ and for random packing of equal-sized spheres $\phi_p = 0.64$. For polydisperse systems, ϕ_p reaches higher values, as is illustrated in Figure 7.3 for one-, two-, three- and four-sized suspensions.

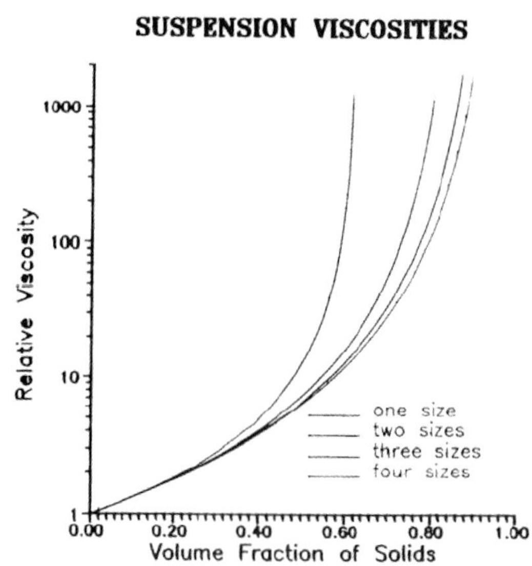

Figure 7.3 Viscosity–volume fraction curves for polydisperse suspensions.

7.1.6
Analysis of the Viscosity-Volume Fraction Curve

The best analysis of the η_r–ϕ curve is due to Krieger and Dougherty [3, 4], who used a mean field approximation by calculating the increase in viscosity as small increments of the suspension are consecutively added. Each added increment corresponds to replacement of the medium by more particles.

They arrived at the following simple semiempirical equation that could fit the viscosity data over the whole volume fraction range:

$$\eta_r = \left(1 - \frac{\phi}{\phi_p}\right)^{-[\eta]\phi_p} \tag{5}$$

Equation 5 is referred to as the Dougherty–Krieger equation [3] and is commonly used for the analysis of the viscosity data.

7.1.7
Rheology of Systems with 'Soft' or Electrostatic Interaction

In this case, the rheology is determined by the double layer repulsion, particularly with small particles and extended double layers [5]. In the low shear rate regime, the viscosity is determined by the Brownian diffusion and the particles approach each other to a distance of the order of $\sim 4.5\kappa^{-1}$ (where κ^{-1} is the 'double layer thickness' that is determined by electrolyte concentration and valency). This means that the effective radius of the particles R_{eff} is much higher than the core radius R. For example, for 100 nm particles with a zeta potential ζ of 50 mV dispersed in a medium of 10^{-5} mol dm^{-3} NaCl ($\kappa^{-1} = 100$ nm), $R_{\text{eff}} \approx 325$ nm.

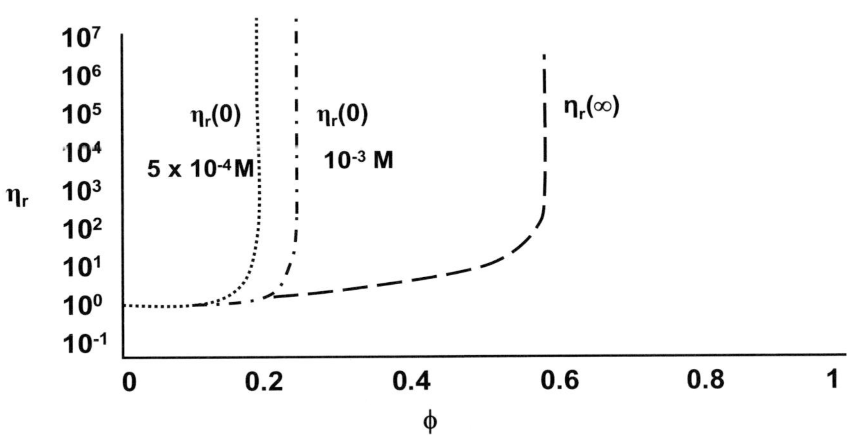

Figure 7.4 Variation of η_r with ϕ for polystyrene latex dispersions at two NaCl concentrations.

The effective volume fraction ϕ_{eff} is also much higher than the core volume fraction. This results in a rapid increase in the viscosity at low core volume fraction [5]. This is illustrated in Figure 7.4, which shows the variation of η_r with ϕ at 5×10^{-4} and 10^{-3} mol dm^{-3} NaCl ($R = 85$ nm and $\zeta = 78$ mV).

The low shear viscosity $\eta_r(0)$ shows a rapid increase at $\phi \approx 0.2$ (the increase occurs at a higher volume fraction at higher electrolyte concentration). At $\phi > 0.2$, the system shows 'solid-like' behavior with $\eta_r(0)$ reaching very high values ($>10^7$). At such high ϕ values, the system shows near plastic flow.

In the high shear rate regime, the increase in η_r occurs at much higher ϕ values. This is illustrated from the plot of the high shear relative viscosity $\eta_r(\infty)$ versus ϕ. At such high shear rates, hydrodynamic interaction predominates over the Brownian diffusion and the system shows a low viscosity denoted by $\eta_r(\infty)$. However ,when ϕ reaches a critical value, pseudoplastic flow is observed.

7.1.8
Viscoelastic Behavior of Electrostatically Stabilized Suspensions

One of the powerful techniques for studying the interaction in electrostatically stabilized suspensions is to use dynamic (oscillatory) measurements [6, 7]. This is illustrated in Figure 7.5, which shows the variation of the complex modulus G^*, storage (elastic) modulus G' and loss (viscous) modulus G'' versus core latex (with radius of 700 nm) volume fraction ϕ. The moduli were measured at low strain (in the linear region) and high frequency (1 Hz).

The trends obtained depend on the NaCl concentration. In 10^{-5} mol dm^{-3} NaCl, the moduli values increase very rapidly at $\phi > 0.46$ and $G' \gg G''$. In 10^{-3}

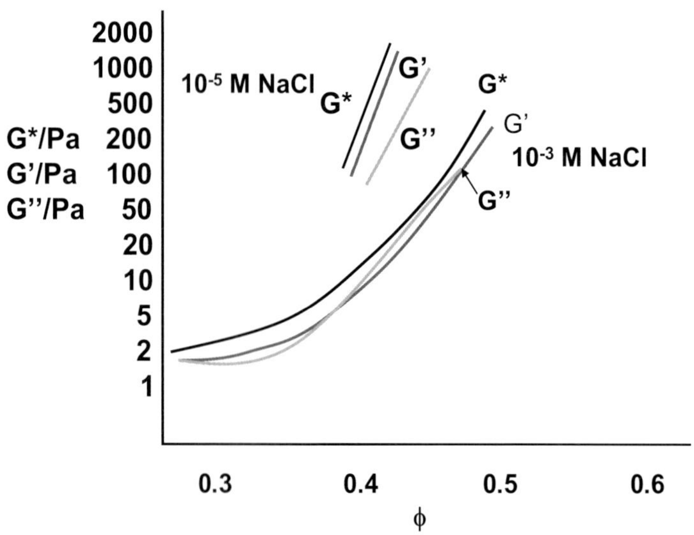

Figure 7.5 Variation of G^*, G' and G'' with ϕ for electrostatically stabilized latex dispersions.

mol dm^{-3} NaCl, the moduli values show a rapid increase at $\phi > 0.57$ and G'' is either lower than or equal to G'.

The above trend reflects the larger ϕ_{eff} values at low NaCl concentration (with a more extended double layer) than at higher NaCl concentration (with a smaller extension).

7.1.9 Rheology of Sterically Stabilized Dispersions

The rheology of sterically stabilized dispersions is determined by the steric repulsion, particularly for small particles with 'thick' adsorbed layers [8]. This is illustrated in Figure 7.6, which shows the variation of G^*, G' and G'' with frequency (Hz) for polystyrene latex dispersions of 175 nm radius containing grafted poly(ethylene oxide) (PEO) with a molecular weight of 2000 (giving a hydrodynamic thickness $\delta \approx 20$ nm). The results clearly show the transition from a predominantly viscous response when $\phi \leq 0.465$ to a predominantly elastic response when $\phi \geq 0.5$. This behavior reflects the steric interaction between the PEO layers. When the surface-to-surface distance between the particles h becomes $<2\delta$, elastic interaction occurs and $G' > G''$.

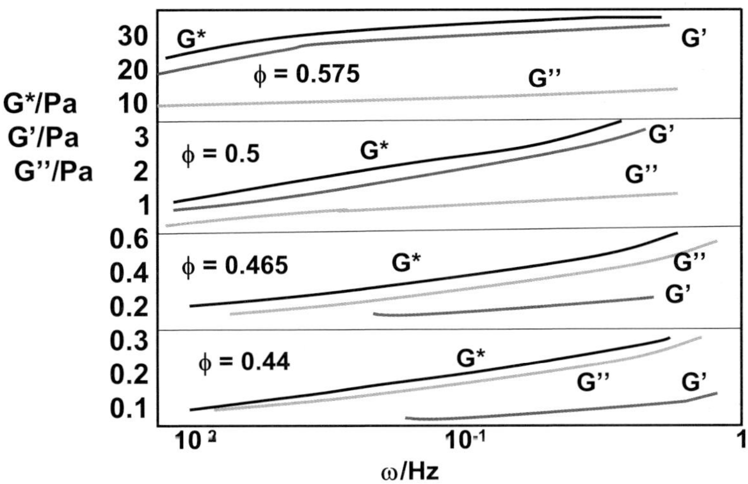

Figure 7.6 Variation of G^*, G' and G'' with frequency for sterically stabilized suspensions.

7.1.10 Rheology of Flocculated Suspensions

These are systems where the net interaction is attractive. As mentioned above, one can distinguish two cases. (i) *Weakly flocculated suspensions:* The attraction in this case is weak (energy of a few kT units) and reversible, e.g. in the secondary minimum of the DLVO (Deryaguin–Landau–Verwey–Overbeek energy of interaction) curve or the shallow minimum obtained with sterically stabilized systems. A particular case of weak flocculation is that obtained on the addition of 'free' (non-adsorbing) polymer referred to as depletion flocculation. (ii) *Strongly flocculated (coagulated) suspensions:* The attraction in this case is strong (involving energies of several hundred kT units) and irreversible.

Study of the rheology of flocculated suspensions is difficult since the structure of the flocs is at non-equilibrium. Theories of flocculated suspensions are also qualitative and based on a number of assumptions.

7.1.10.1 Weakly Flocculated Suspensions

As mentioned in Chapter 4, weak flocculation may be obtained the addition of 'free' (non-adsorbing) polymer to a sterically stabilized dispersion [9]. This is exemplified by a latex dispersion (of radius of 175 nm) containing grafted PEO chains of molecular weight $M = 2000$ to which 'free' PEO is added at various concentrations. PEOs with three molecular weights were used: $M = 20\,000$, $35\,000$ and $90\,000$. As an illustration, Figure 7.7 shows the variation of the Bingham yield value σ_β, Casson yield value σ_C and shear modulus G_∞ with volume fraction of 'free' polymer ϕ_p (PEO, $M = 20\,000$).

All rheological parameters show a rapid increase at and above a critical ϕ_p (denoted $\phi_p{}^+$) of 0.02, indicating the onset of weak flocculation. Similar trends are obtained using oscillatory measurements as shown in Figure 7.8. The value of $\phi_p{}^+$ decreases with increase in M: $\phi_p{}^+ = 0.01$ for $M = 35\,000$ and $\phi_p{}^+ = 0.005$ for $M = 90\,000$.

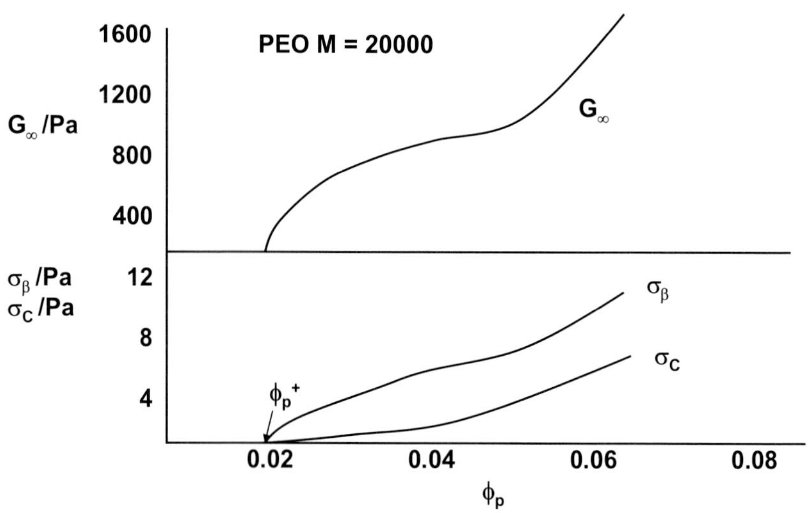

Figure 7.7 Variation of yield value and shear modulus with 'free' polymer volume fraction (PEO, $M = 20\,000$).

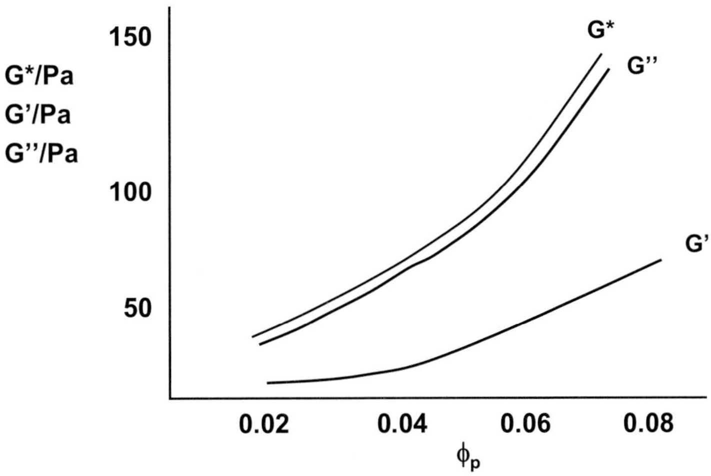

Figure 7.8 Variation of G^*, G' and G'' with ϕ_p (PEO, $M = 20\,000$).

7.1.10.2 **Strongly Flocculated (Coagulated) Suspensions**

Steady-state shear stress–shear rate curves show a pseudoplastic flow curve as illustrated in Figure 7.9. The flow curve is characterized by three main parameters: (i) the shear rate above which the flow curve shows linear behavior – above this shear rate, collisions occur between the flocs, which may cause interchange between the flocculi (the smaller floc units that aggregate to form a floc); in this linear region, the ratio of the floc volume to the particle volume (ϕ_F/ϕ_p), i.e. the floc density, remains constant; (ii) σ_β, the residual stress (yield stress) that arises from the residual effect of inter-particle potential; and (iii) η_{pl}, the slope of the linear portion of the flow curve that arises from purely hydrodynamic effects.

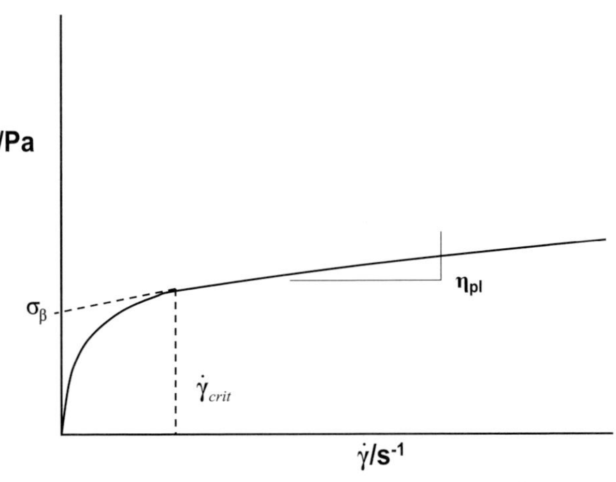

Figure 7.9 Pseudoplastic flow curve for a flocculated suspension.

7.1.11
Analysis of the Flow Curve

7.1.11.1 Impulse Theory: Goodeve and Gillespie [10, 11]

The interparticle interaction effects (given by σ_β) and hydrodynamic effects (give by η_{pl}) are assumed to be additive:

$$\sigma = \sigma_\beta + \eta_{pl}\dot{\gamma} \tag{6}$$

To calculate σ_β, Goodeve [10] proposed that when shearing occurs, links between particles in a flocculated structure are stretched, broken and re-formed. An impulse is transferred from a fast-moving layer to a slow-moving layer. Non-Newtonian effects are due to effect of shear on the number of links, the lifetime of a link and any change in the size of the floc.

According to the Goodeve theory, the yield value is given by

$$\sigma_\beta = \left(\frac{3\phi^2}{2\pi a^3}\right) E_A \tag{7}$$

where ϕ is the volume fraction of the dispersed phase, a is the particle radius and E_A is the total binding energy:

$$E_A = n_L \varepsilon_L \tag{8}$$

where n_L is the number of links with a binding energy ε_L per link.

According to equation 7, $\sigma_\beta \propto\propto \phi^2$, $\propto (1/a^3)$ and $\propto E_A$ (the energy of attraction).

7.1.11.2 Elastic Floc Model: Hunter and Co-workers [12, 13]

The floc is assumed to consist of an open network of 'girders' as shown schematically in Figure 7.10. The floc undergoes extension and compression during rotation in a shear flow. The bonds are stretched by a small amount Δ (that can be as small as 1% of the particle radius).

To calculate σ_β, Hunter and co-workers considered the energy dissipation during rupture of the flocs (assumed to consist of doublets). The yield value σ_β is given by the expression

$$\sigma_\beta = \alpha_0 \beta \lambda \eta \dot{\gamma} \left(\frac{{a_{floc}}^2}{a^3}\right) {\phi_s}^2 \Delta C_{FP} \tag{9}$$

where α_0 is the collision frequency, β is a constant ($= 27/5$), λ is a correction factor (~ 1) and C_{FP} is the floc density ($= \phi_F/\phi_s$).

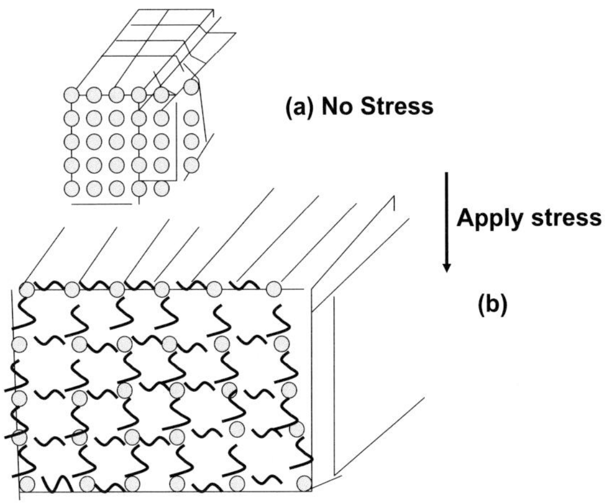

Figure 7.10 Schematic diagram of the elastic floc.

7.1.12 Fractal Concept of Flocculation

The floc structure can be treated as fractals whereby an isolated floc with radius a_F can be assumed to have uniform packing throughout that floc [14, 15].

In the above case, the number of particles in a floc is given by

$$n_f = \phi_{mf}\left(\frac{a_F}{a}\right)^3 \tag{10}$$

where ϕ_{mf} is the packing fraction of the floc.

If the floc does not have constant packing throughout its structure, but is dendritic in form, the packing density of the floc begins to reduce as one goes from the center to the edge. If this reduction is with a constant power law D,

$$n_F = \left(\frac{a_F}{a}\right)^D \tag{11}$$

where $0 < D \leq 3$.

D is called the packing index and it represents the packing change with distance from the center.

In rapid aggregation (diffusion-limited aggregation, DLA), when particles touch they stick. Particle–particle aggregation gives $D = 2.5$; aggregate–aggregate aggregation gives $D = 1.8$. The lower the value of D, the more open the floc structure is.

Hence by determining D one can obtain information on the flocculation behavior. If flocculation of a suspension occurs by changing the conditions (e.g. increasing temperature), one can visualize sites for nucleation of flocs occurring randomly throughout the whole volume of the suspension.

The total number of primary particles does not change and the volume fraction of the floc is given by

$$\phi_{\mathrm{F}} = \phi \left(\frac{a_{\mathrm{F}}}{F} \right)^{3-D} \tag{12}$$

Since the yield stress σ_β and elastic modulus G' depend on the volume fraction, one can use a power law in the form

$$\sigma_\beta = K\phi^m \tag{13}$$

$$G' = K\phi^m \tag{14}$$

where the exponent m reflects the fractal dimension.

Thus, by plotting log σ_β or log G' versus log ϕ, one can obtain m from the slope and this can be used to characterize the nature and structure of the floc.

7.2 Rheology of Emulsions

7.2.1 Introduction

The rheology of emulsions has many similar features to those of suspensions. However, they differ in three main respects: (i) the mobile liquid/liquid interface that contains surfactant or polymer layers introduces a response to deformation and one has to consider the interfacial rheology; (ii) the dispersed phase viscosity relative to that of the medium has an effect on the rheology of the emulsion; and (iii) the deformable nature of the disperse phase droplets, particularly for large droplets, has an effect on the emulsion rheology at high phase volume fraction ϕ.

When the above factors are considered, one can treat the bulk rheology of emulsions in a similar way as for suspensions and the same techniques can be applied.

7.2.2 Interfacial Rheology

7.2.2.1 Interfacial Tension and Surface Pressure

A fluid interface in equilibrium exhibits an intrinsic state of tension that is characterized by its interfacial tension γ, which is given by the change in free energy with area of the interface, at constant composition n_i and temperature T:

$$\gamma = \left(\frac{\partial G}{\partial A}\right)_{n_i, T} \tag{15}$$

The unit for γ is energy per unit area (mJ m^{-1}) or force per unit length (mN m^{-1}), which are dimensionally equivalent.

Adsorption of surfactants or polymers lowers the interfacial tension, which produces a two-dimensional surface pressure π that is given by

$$\pi = \gamma_0 - \gamma \tag{16}$$

where γ_0 is the interfacial tension of the 'clean' interface (before adsorption) and γ that after adsorption.

7.2.2.2 Interfacial Shear Viscosity

The interface is considered to be a macroscopically planer, dynamic fluid interface. Thus, the interface is regarded as a two-dimensional entity independent of the surrounding three-dimensional fluid. The interface is considered to correspond to a highly viscous insoluble monolayer and the interfacial stress σ_s acting within such a monolayer is sufficiently large compared with the bulk fluid stress acting across the interface, and in this way one can define an interfacial shear viscosity η_s:

$$\sigma_s = \eta_s \dot{\gamma} \tag{17}$$

where $\dot{\gamma}$ is the shear rate, and η_s is given in surface Pa s (N m^{-1} s) or surface poise (dyne cm^{-1} s).

It should be mentioned that the surface viscosity of a surfactant-free interface is negligible and it can reach high values for adsorbed rigid molecules such as proteins.

7.2.2.3 Measurement of Interfacial Viscosity

Many surface viscometers utilize torsional stress measurements upon a rotating a ring, disk or knife edge (shown schematically in Figure 7.11) within or near to the

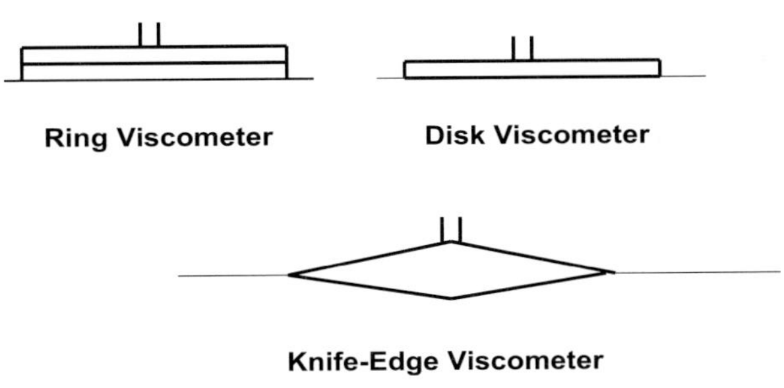

Figure 7.11 Schematic representation of surface viscometers.

liquid/liquid interface [16]. This type of viscometer is moderately sensitive; for a disk viscometer, the interfacial shear viscosity can be measured in the range $\eta_s \geq 10^{-2}$ surface Pa s.

The disk is rotated within the plane of the interface with angular velocity ω. A torque is exerted upon the disk of radius R by both the surfactant film with surface viscosity η_s and the viscous liquid (with bulk viscosity η) that is given by

$$M = \frac{8}{3} R^3 \eta \omega + 4\pi R^2 \eta_s \omega \tag{18}$$

7.2.3
Interfacial Dilational Elasticity

The interfacial dilational (Gibbs) elasticity ε, which is an important parameter in determining emulsion stability (reduction of coalescence during formation), is given by

$$\varepsilon = \frac{d\gamma}{d \ln A} \tag{19}$$

where $d\eta$ is the change in interfacial tension during expansion of the interface by an amount dA (referred to as interfacial tension gradient resulting from non-uniform surfactant adsorption on expansion of the interface).

One of the most convenient methods for the measurement of ε is to use a Langmuir trough with two moving barriers for expansion and compression of the interface. Another method is to use the oscillating bubble technique and instruments are commercially available.

A useful method for the measurement of ε is the pulsed drop method [17]. Rapid expansion of a droplet at the end of a capillary from a radius r_1 to r_2 is ob-

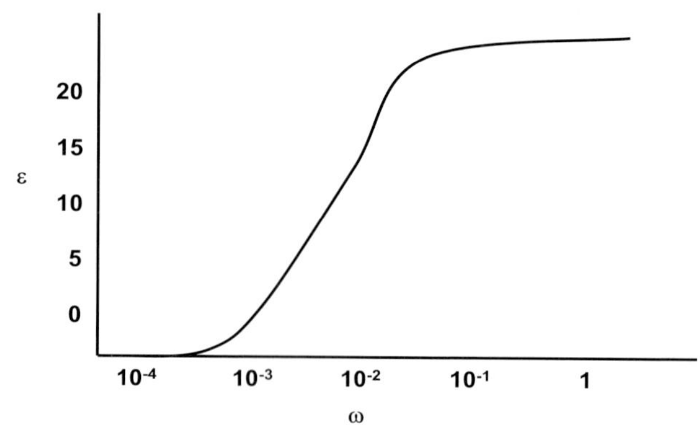

Figure 7.12 Gibbs dilational elasticity versus frequency.

tained by application of pressure. The pressure drop within the droplet is measured as a function of time using a sensitive pressure transducer. From the pressure drop, one can obtain the interfacial tension as a function of time. The Gibbs dilational elasticity is determined from values of the time-dependent interfacial tension. Measurement can be made as a function of frequency, as illustrated in Figure 7.12 for stearic acid at the decane/water interface at pH 2.5

7.2.4
Interfacial Dilational Viscosity

Measurement of the dilational viscosity is more difficult than measurement of the interfacial shear viscosity, due to the coupling between dilational viscous and elastic components. The most convenient method for the measurement of dilational viscosity is the maximum bubble pressure technique, which can be only applied at the air/water interface. According to this technique, the pressure drop across the bubble surface at the instant when the bubble possesses a hemispherical shape (corresponding to the maximum pressure) is due to a combination of bulk viscous, surface tension and surface dilational viscosity effects, and this allows one to obtain the interfacial dilational viscosity.

7.2.5
Non-Newtonian Effects

Most adsorbed surfactant and polymer coils at the oil/water (O/W) interface show non-Newtonian rheological behavior. The surface shear viscosity η_s depends on the applied shear rate, showing shear thinning at high shear rates. Some films also show Bingham plastic behavior with a measurable yield stress.

Many adsorbed polymers and proteins show viscoelastic behavior and one can measure viscous and elastic components using sinusoidally oscillating surface dilation. For example, the complex dilational modulus ε^* obtained can be split into an 'in-phase' (the elastic component ε') and 'out-of-phase' (the viscous component ε'') components. Creep and stress relaxation methods can be applied to study viscoelasticity.

7.2.6
Correlation of Emulsion Stability with Interfacial Rheology

7.2.6.1 Mixed Surfactant Films

Prins *et al.* [18] found that a mixture of sodium dodecyl sulfate (SDS) and dodecyl alcohol gives a more stable O/W emulsion compared with emulsions prepared using SDS alone. This enhanced stability is due to the higher interfacial dilational elasticity ε for the mixture compared with that of SDS alone. Interfacial dilational viscosity did not play a major role since the emulsions are stable at high temperature, whereby the interfacial viscosity becomes lower.

The above correlation is not general for all surfactant films since other factors, such as thinning of the film between emulsion droplets (which depends on other factors such as repulsive forces), can also play a major role.

7.2.6.2 **Protein Films**

Biswas and Haydon [19] found some correlation between the viscoelastic properties of protein (albumin or arabinic acid) films at the O/W interface and the stability of emulsion drops against coalescence. Viscoelastic measurements were carried out using creep and stress relaxation measurements (using a specially designed interfacial rheometer). A constant torque or stress σ (mN m^{-1}) was applied and the deformation γ was measured as a function of time for 30 min. After this period, the torque was removed and γ (which changes sign) was measured as a function of time to obtain the recovery curve. The results are illustrated in Figure 7.13.

From the creep curves, one can obtain the instantaneous modulus G_0 $(\sigma/\gamma_{\text{int}})$ and the surface viscosity η_s from the slope of the straight line (which gives the shear rate) and the applied stress. G_0 and η_s are plotted versus pH as shown in Figure 7.14. Both show an increase with increase in pH, reaching a maximum at pH $\approx$ 6 (the isoelectric point of the protein), at which the protein molecules show maximum rigidity at the interface.

The stability of the emulsion was assessed by measuring the residence time t of several oil droplets at a planer O/W interface containing the adsorbed protein. Figure 7.14 shows the variation of $t_{\frac{1}{2}}$ (time taken for half the number of oil droplets to coalesce with the oil at the O/W interface) with pH. A good correlation between $t_{\frac{1}{2}}$, G_0 and η_s is obtained.

Biswas and Haydon [19] derived a relationship between coalescence time τ and surface viscosity η_s, instantaneous modulus G_0, and adsorbed film thickness h:

$$\tau = \eta_s \left[3C' \frac{h^2}{A} - \frac{1}{G_0} - \phi(t) \right] \tag{20}$$

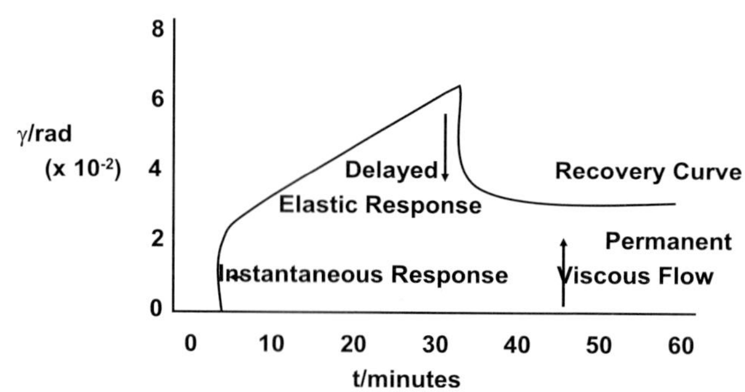

Figure 7.13 Creep curve for a protein film at the O/W interface.

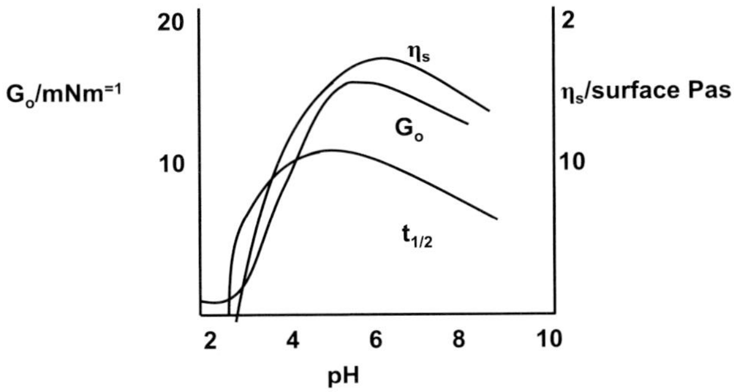

Figure 7.14 Variation of $t_{\frac{1}{2}}$, G_0 and η_s with pH.

where $3C'$ is a critical deformation factor, A is the Hamaker constant and $\phi(t)$ is the elastic deformation per unit stress.

Equation 20 shows that τ increases with increase in η_s, but most importantly it is directly proportional to h^2. These results show that viscoelasticity is necessary but not sufficient to ensure stability against coalescence. To ensure stability of an emulsion, one must ensure that h is large enough and film drainage is prevented.

7.2.7
Bulk Rheology of Emulsions

For rigid (highly viscous) oil droplets dispersed in a medium of low viscosity such as water, the relative viscosity η_r of a dilute (volume fraction $\phi \leq 0.01$) O/W emulsion of non-interacting droplets behaves as 'hard spheres' (similar to suspensions).

In the above case, η_r is given by the Einstein equation:

$$\eta_r = 1 + [\eta]\phi \tag{21}$$

where $[\eta]$ is the intrinsic viscosity, which is equal to 2.5 for hard spheres.

For droplets with low viscosity (comparable to that of the medium), the transmission of tangential stress across the O/W interface, from the continuous phase to the dispersed phase, causes liquid circulation in the droplets. Energy dissipation is less than that for hard spheres and the relative viscosity is lower than that predicted by the Einstein equation.

For an emulsion with viscosity η_i for the disperse phase and η_0 for the continuous phase,

$$[\eta] = 2.5\left(\frac{\eta_i + 0.4\eta_0}{\eta_i + \eta_0}\right) \tag{22}$$

Clearly, when $\eta_i \gg \eta_0$, the droplets behave as rigid spheres and $[\eta]$ approaches the Einstein limit of 2.5. In contrast, if $\eta_i \ll \eta_0$ (as is the case for foams), $[\eta] = 1$.

In the presence of viscous interfacial layers, equation 22 is modified to take into account the surface shear viscosity η_s and surface dilational viscosity μ_s:

$$[\eta] = 2.5\left(\frac{\eta_i + 0.4\eta_0 + \xi}{\eta_i + \eta_0 + \xi}\right) \tag{23}$$

where

$$\xi = \frac{2\eta_s + 3\mu_s}{R} \tag{24}$$

and R is the droplet radius.

When the volume fraction of droplets exceeds the Einstein limit, i.e. $\phi > 0.01$, one must take into account the effect of Brownian motion and interparticle interactions. The smaller the emulsion droplets, the more important is the contribution of Brownian motion and colloidal interactions. Brownian diffusion tends to randomize the position of colloidal particles, leading to the formation of temporary doublets, triplets, etc. The hydrodynamic interactions are of longer range than the colloidal interactions and they come into play at relatively low volume fractions ($\phi > 0.01$), resulting in ordering of the particles into layers and tending to destroy the temporary aggregates caused by the Brownian diffusion. This explains the shear thinning behavior of emulsions at high shear rates.

For the volume fraction range $0.01 < \phi < 0.2$, Batchelor [2] derived the following expression for a dispersion of hydrodynamically interacting hard spheres:

$$\eta_r = 1 + 2.5\phi + 6.2\phi^2 + \vartheta\phi^3 \tag{25}$$

The second term on the right-hand side in equation 25 is the Einstein limit, the third term accounts for hydrodynamic (two-body) interaction and the fourth term relates to multi-body interaction.

At higher volume fractions ($\phi > 0.2$), η_r is a complex function of ϕ and the η_r–ϕ curve is shown schematically in Figure 7.15. This curve is characterized by two asymptotes, $[\eta]$ the intrinsic viscosity and ϕ_p the maximum packing fraction.

A good semiempirical equation that fits the curve was given by Krieger and Dougherty [3, 4]:

$$\eta_r = \left(1 - \frac{\phi}{\phi_p}\right)^{-[\eta]\phi_p} \tag{26}$$

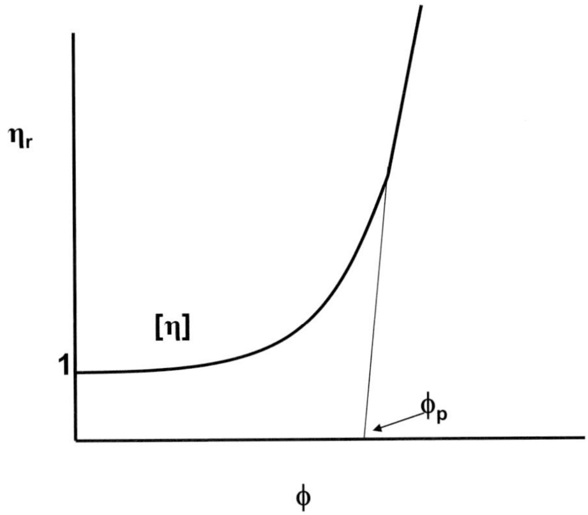

Figure 7.15 η_r–ϕ curve.

7.2.8
Experimental η_r–ϕ Curves

Experimental results for η_r–ϕ curves were obtained for paraffin O/W emulsions [20] stabilized with an A–B–C surfactant consisting of nonylphenol (B), 13 mol of propylene oxide (C) and poly(ethylene oxide) (PEO) with 27, 48, 80 and 174 mol of EO. As an illustration, Figure 7.16 shows the results for an emulsion stabilized with the surfactant containing 27 EO (the volume medium diameter of the droplets is 3.5 μm). The calculations based on the Dougherty–Krieger

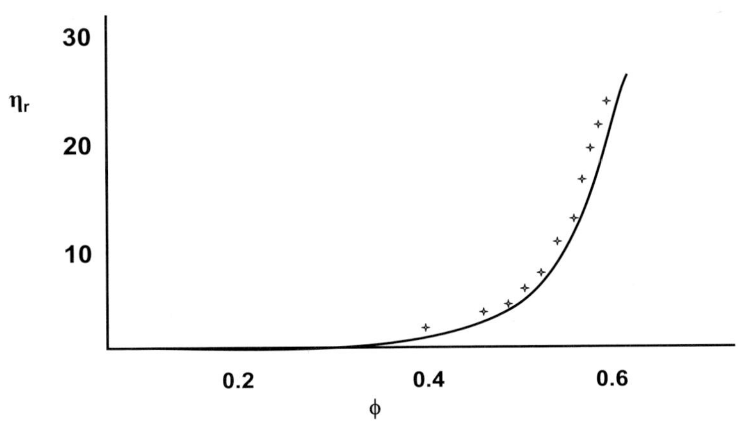

Figure 7.16 Experimental and theoretical η_r–ϕ curves. +, Experimental results; line, calculated using the Dougherty–Krieger equation.

equation are also shown. In these calculations, $[\eta] = 2.5$ and ϕ_p was obtained from a plot of $\eta^{-1/2}$ versus ϕ and extrapolation of the straight line to $\eta^{-\frac{1}{2}} = 0$. The value of ϕ_p was 0.73 (which is higher than the maximum random packing of 0.64, as a result of the polydispersity of the emulsion). The results using the other three surfactants showed the same trend; the experimental η_r–ϕ curves are close to those calculated using the Dougherty–Krieger equation, indicating that these emulsions behave as hard spheres.

7.2.8.1 Influence of Droplet Deformability

The influence of droplet deformability on emulsion rheology was investigated by Saiki and Prestidge [21] by comparing the η_r–ϕ curves for hard spheres of silica with two polydimethylsilioxane (PDMS) emulsions with low (PDMS 0.3) and high deformability (PDMS 0.45) (by controlling the proportion of cross-linking agent for the droplets; 0.3 low and 0.45 high cross-linking agent). The η_r–ϕ curves for the three systems are shown in Figure 7.17. The η_r–ϕ curve for silica can be fitted by the Dougherty–Krieger equation over the whole volume fraction range, indicating typical hard-sphere behavior. The η_r–ϕ curve for the less deformable PDMS deviates from the hard-sphere curve at $\phi = 0.58$. The η_r–ϕ curve for the more deformable PDMS deviates from the hard-sphere curve at $\phi = 0.40$. This clearly shows that the deformation of the 'soft' droplets occurs at relatively low volume fraction.

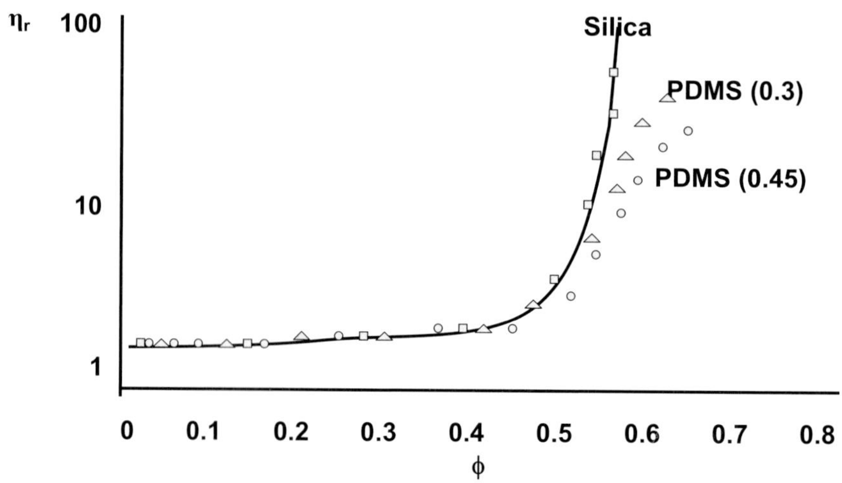

Figure 7.17 η_r–ϕ curves for silica and two PDMS emulsions.

7.2.9 Viscoelastic Properties of Concentrated Emulsions

The viscoelastic properties of emulsions can be investigated using dynamic (oscillatory) measurements. A sinusoidal strain with amplitude γ_0 is applied on the system at a frequency ω (rad s^{-1}) and the stress σ (with amplitude σ_0) is simultaneously measured. From the time shift Δt between the sine waves of strain and stress, one can measure the phase angle shift δ ($\delta = \Delta t\omega$).

From σ_0, γ_0 and δ, one can obtain the complex modulus G^*, the storage modulus G' (the elastic component) and the loss modulus G'' (the viscous component). G^*, G' and G'' are measured as a function of strain amplitude to obtain the linear viscoelastic region and then as a function of frequency (keeping γ_0 in the linear region). As an illustration, Figure 7.18 shows the results for an O/W emulsion at $\phi = 0.6$ (the emulsion was prepared using an A–B–A block copolymer of poly(ethylene oxide) (PEO, A) and poly(propylene oxide) (PPO, B) with an average of 47 PO units and 42 EO units [22].

The results in Figure 7.18 are typical for a viscoelastic liquid. In the low-frequency regime (<1 Hz), $G'' > G'$. As the frequency ω increases, G' increases and at a characteristic frequency ω^* (the cross-over point) G' becomes higher than G'' and at high frequency it becomes closer to G^*. G'' increases with increase in frequency, reaching a maximum at ω^*, after which it decreases with further increase in frequency.

From ω^*, one can calculate the relaxation time t^*:

$$t^* = \frac{1}{2\pi\omega^*} \tag{27}$$

For the above value of ϕ (= 0.6), $t^* = 0.12$ s; t^* increases with increase in ϕ and this reflects the stronger interaction with increase of ϕ.

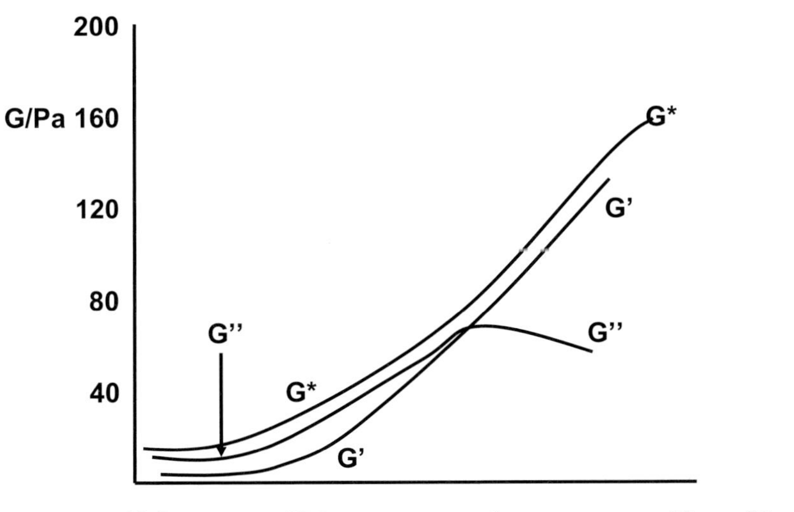

Figure 7.18 Variation of G^*, G' and G'' with frequency ω.

To obtain the onset of strong elastic interaction in emulsions, G^*, G' and G'' (obtained in the linear viscoelastic region and at high frequency, e.g. 1 Hz) are plotted versus the volume fraction of the emulsion ϕ. One should ensure that the droplet size distribution in all emulsions is the same. The most convenient way is to prepare an emulsion at the highest possible ϕ (e.g. 0.6), then this emulsion is diluted to obtain various ϕ values. Droplet size analysis should be performed for each emulsion to ensure that the size distribution is the same.

Figure 7.19 shows the plots for G^*, G' and G'' versus ϕ. At $\phi < 0.56$, $G'' > G'$, whereas at $\phi > 0.56$, $G' > G''$; $\phi = 0.56$ is the onset of predominantly elastic interaction and this reflects the small distance of separation between the droplets.

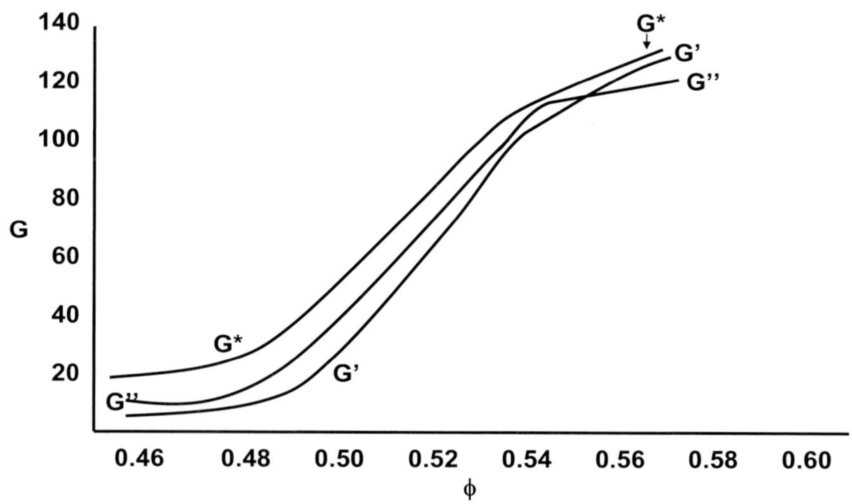

Figure 7.19 Variation of G^*, G' and G'' with ϕ.

7.3 Rheology of Suspoemulsions of Polystyrene Latex and Isoparaffinic Oil Stabilized with Synperonic PE (PEO–PPO–PEO A–B–A Block Copolymer)

As an illustration, Figure 7.20 shows typical plots of G^* (complex modulus), G' (storage or elastic modulus), G'' (loss or viscous modulus) and η' (dynamic viscosity) as a function of frequency (Hz) for 90% emulsion and 10% latex both stabilized with Synperonic PE block copolymer [23, 24]. The results shown were obtained at low strain amplitudes (i.e. in the linear viscoelastic region). Both G^* and G' show a rapid increase with increase in frequency above 0.1 Hz. However, G'', which is higher than G' at frequencies below 1 Hz, starts to decrease with increase in frequency above 1 Hz. The dynamic viscosity shows a decrease with increase in frequency, as expected (shear thinning system).

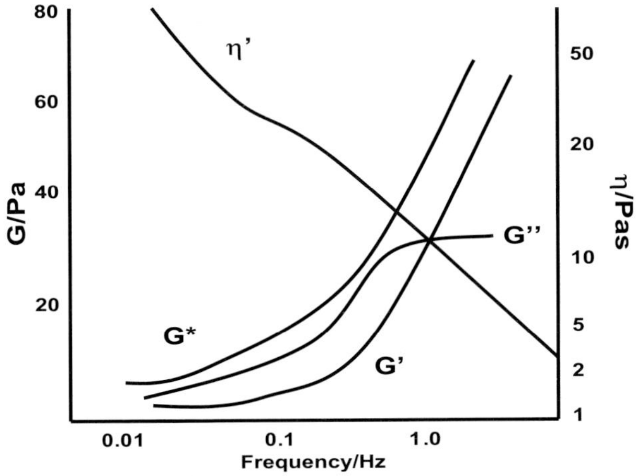

Figure 7.20 Variation of G^*, G', G'' and η' with frequency for a suspoemulsion of 90% isoparaffinic O/W emulsion and 10% polystyrene latex.

A well-defined cross-over point ($G' = G''$) can be identified from which the relaxation time of the system can be calculated:

$$\tau = \frac{1}{\omega^*} \tag{28}$$

where * is the characteristic frequency (rad s^{-1}) at the cross-over point. Note that $\omega = 2\pi\nu$, where ν is the frequency in hertz.

Similar results were obtained for other suspoemulsions with various emulsion:latex ratios. The same trend was also obtained for the emulsion and latex dispersions alone. Addition of the latex to the emulsion causes a shift in τ values to higher frequencies, indicating a stronger interaction between the latex particles and emulsion droplets.

Figure 7.21 shows the variation of G^*, G' and G'' at $\phi = 0.57$ and $\nu = 1$ Hz with percentage of emulsion and latex in the suspoemulsion. The emulsion has much higher moduli than the latex at the same volume fraction. Although the emulsion has a VMD (Volume Mean Diameter) (0.98 μm) that is close to the latex (1.18 μm), the former is much more polydisperse than the latter. The much smaller emulsion droplets present may account for the higher moduli of the emulsion compared with the latex. As the proportion of the latex in the suspoemulsion is increased, the moduli decrease. Replacement of emulsion with latex would mean replacing a proportion of the smaller emulsion droplets with larger latex particles, which results in a reduction in the moduli values. It should be mentioned, however, that the mixture of emulsion and latex becomes relatively more elastic than viscous, indicating stronger interaction between the emulsion droplets and the latex particles.

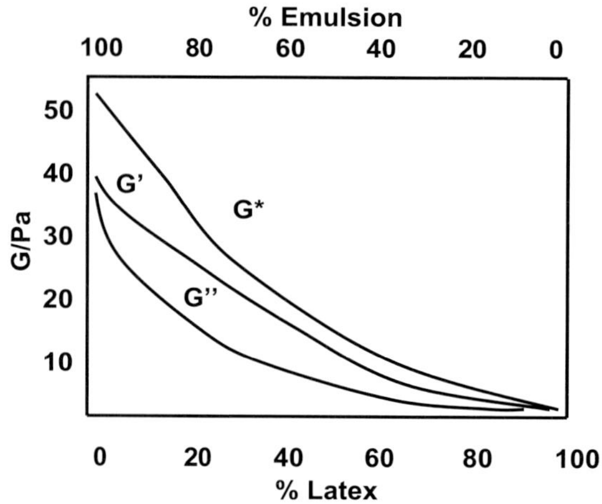

Figure 7.21 Variation of G^*, G' and G'' with percentage of emulsion and latex; $\phi = 0$.

7.3.1
Model Systems of Polystyrene Latex with Grafted PEO Chains and Hexadecane Emulsions

It is clear from the above discussion that the interaction between suspension particles and emulsion droplets depends on the nature of the stabilizer used for the particles and droplets. For that reason, we investigated model systems in which the latex particles contained grafted PEO chains (with no possible desorption) and the emulsion was based on hexadecane stabilized with Pluronic PE L92 (containing 20% PEO). The particle and droplet radii were very similar (315 and 280 nm, respectively) in order to avoid complications arising from the change in particle size distribution on mixing the suspension and emulsion.

Steady-state shear stress–shear rate curves were used to obtain the relative viscosity (η_r)–volume fraction (ϕ) relationship for the latex and emulsion. The results are shown in Figure 7.22, which also contains the theoretically predicted curve based on the Dougherty–Krieger equation 5, where ϕ_p is the maximum packing fraction and $[\eta]$ is the intrinsic viscosity, which is equal to 2.5 for hard spheres. Two values for ϕ_p were used, namely 0.60 and 0.61 for the emulsion and latex, respectively. Reasonable agreement between the experimental η_r–ϕ curve and the theoretical curve based on equation 5 for the latex dispersions was obtained, indicating that this system behaves as near hard spheres. However, the results for the emulsion deviate from the theoretical curves since the emulsion droplets are deformable.

Figure 7.23 shows log–log plots of the elastic modulus G' (measured at low strain amplitudes to ensure the linear viscoelastic region and a frequency of 1 Hz) versus volume fraction of the latex, emulsion and various mixtures of emulsion–latex systems. All results are similar for volume fractions below

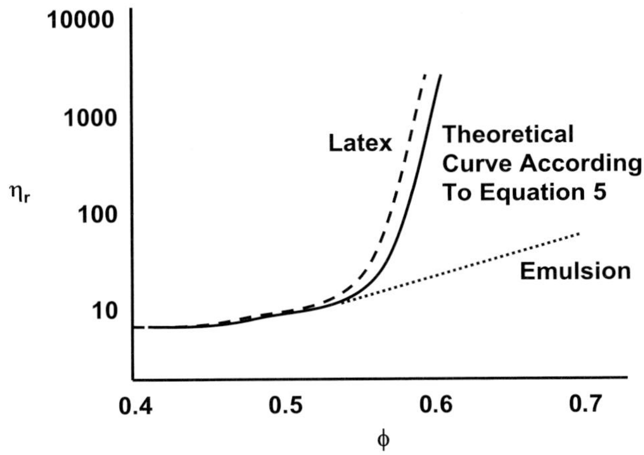

Figure 7.22 η_r–ϕ curves for latexes and emulsions.

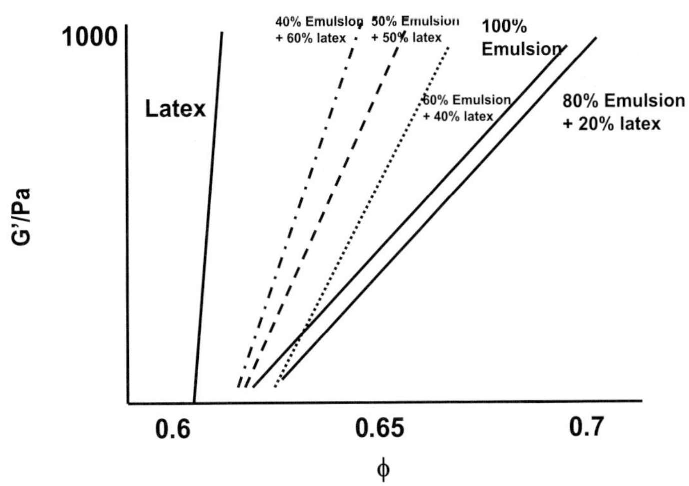

Figure 7.23 Log–log plots of G' versus ϕ for latexes, emulsions and their mixtures.

~0.62, indicating that the interactions between the emulsion droplets and latex particles are of the same nature and both particles and droplets behave as near hard spheres. However, at $\phi > 0.62$, the behaviors of the latexes and emulsions differ significantly, as indicated by the much reduced slope of the log G'–log ϕ curve for the emulsions compared with the latexes. Above this volume fraction, the interaction between the emulsion droplets is fairly high and the system can reduce this interaction by deformation of the emulsion droplets. This situation is not possible with the latexes, where the particles are rigid. Similar behavior is observed for the suspoemulsion when the percentage of the emulsion in the mixture is >60%. This implies that the behavior of emulsions and suspoemulsions with more than 60% emulsion is close in its rheological behavior to concentrated emulsions, and this has implications in the formulation of suspoemulsions.

The G'–ϕ curves can be analyzed using the model suggested by Princen and Kiss [25], who assumed a system of cylinders arranged in a hexagonal array. When such an arrangement is strained, the total interface is increased; this creates a restoring force that is proportional to the interfacial tension γ. Using this model, Princen and Kiss [25] derived an expression relating the shear modulus G_0 (which can be replaced by the high-frequency modulus G') to the volume fraction ϕ:

$$G' = a\left(\frac{\gamma}{R_{32}}\right)\phi^{\frac{1}{3}}(\phi - b) \qquad (29)$$

where a and b are constants equal to 1.7 and 0.71, respectively. The origin of the constant b is the value of the maximum packing of undistorted cylinders in the array. R_{32} is the average volume-to-surface ratio of the radius R (the Sauter radius).

For the latex–emulsion systems studied, R_{32} for the emulsion is constant and, due to the excess surfactant in the emulsion, γ is expected to be little dependent on the composition. The values of the parameters to fit the data of Figure 7.23 are given in Table 7.1

The parameter b corresponds to the volume fraction at the onset of elasticity.

The mixture of latex and emulsion may be regarded as two elastic elements in series with the appropriate volume fractions:

$$\frac{1}{G_m} = \frac{f_e}{G_e} + \frac{1 - f_e}{G_L} \qquad (30)$$

where G_m, G_e and G_L are the elastic moduli of the mixture, emulsion and latex, respectively, and F_e is the weight fraction of the emulsion in the mixture.

Table 7.1 Parameters of the fit of equation 29 for emulsion, latex and their mixtures.

Emulsion (%)	$a\gamma/R_{32}$	b
100	9896 ± 690	0.63 ± 0.01
80	9900 ± 1700	0.63 ± 0.01
60	17 500 ± 2000	0.63 ± 0.03
50	20 700 ± 2200	0.63 ± 0.01
40	23 700 ± 2200	0.62 ± 0.02
20	~100 000	0.61 ± 0.01
0	~1 000 000	0.62 ± 0.02

A plot of the linear lines of Figure 7.23 (which is equal to $a\gamma/R_{32}$) is shown in Figure 7.24 together with the predicted line based on equation 29. The agreement between the values of the slopes and those predicted using equation 29 is good.

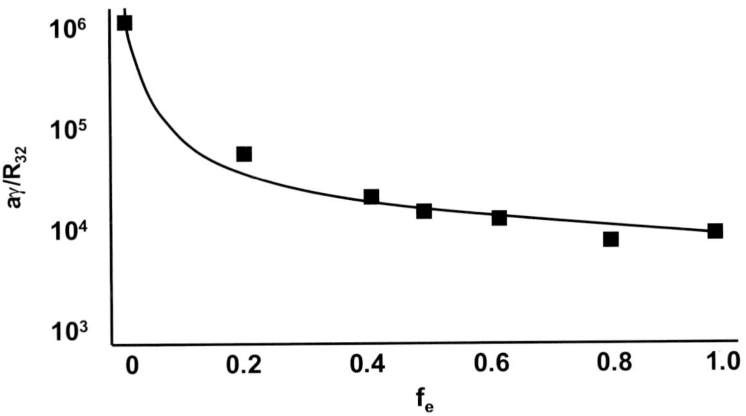

Figure 7.24 $a\gamma/R_{32}$ versus f_e. Line drawn using equation 30.

References

1 A. Einstein, *Ann. Phys.*, *19*, 289 (1906); *34*, 591 (1911).

2 G.K. Batchelor, *J. Fluid Mech.*, *83*, 97 (1977).

3 I.M. Krieger and T.J. Dougherty, *Trans. Soc. Rheol.*, *3*, 137 (1959).

4 I.M. Krieger, *Adv. Colloid Interface Sci.*, *3*, 111 (1972).

5 J.W. Goodwin and R.W. Hughes, *Rheology for Chemists*, Royal Society of Chemistry, Cambridge (2000).

6 Th.F. Tadros, Rheology of concentrated stable and flocculated suspensions, in *Flocculation and Dewatering*, ed. B.M. Mougdil and B.J. Scheiter, Engineering Foundation Publishers, New York, pp. 43–87 (1989).

7 Th.F. Tadros, *Adv. Colloid Interface Sci.*, *68*, 97 (1996).

8 C. Prestidge and Th.F. Tadros, *J. Colloid Interface Sci.*, *124*, 660 (1988).

9 C. Prestidge and Th.F. Tadros, *Colloids Surf.*, *31*, 325 (1988).

10 C.V. Goodeve, *Trans. Faraday Soc.*, *35*, 342 (1939).

11 T. Gillespie, *J. Colloid Sci.*, *15*, 219 (1960).

12 R.J. Hunter and S.K. Nicol, *J. Colloid Interface Sci.*, *28*, 200 (1968).

13 B.A. Firth and R.J. Hunter, *J. Colloid Interface Sci.*, *57*, 248, 257, 266 (1976).

14 P.D.A. Mills, J.W. Goodwin and B. Grover, *Colloid Polym. Sci.*, *269*, 949 (1991).

15 J.W. Goodwin and R.W. Hughes, *Adv. Colloid Interface Sci.*, *42*, 303 (1992).

16 D.W. Criddle, The viscosity and viscoelasticity of interfaces, in *Rheology*, ed. F.R. Eirich, Academic Press, New York, Vol. 3, Chapter 11 (1960).

17 D.A. Edwards, H. Brenner and D.T. Wasan, *Interfacial Transport Processes and Rheology*, Butterworth-Heinemann, Boston (1991).

18 A. Prins, C. Arcuri and M. van den Tempel, *J. Colloid Interface Sci.*, *24*, 811 (1967).

19 B. Biswas and D.A. Haydon, *Proc. R. Soc. London, Ser. A*, *271*, 296 (1963); *2*, 317 (1963); *Kolloid Z.*, *185*, 31 (1962); *186*, 57 (1962).

20 Th.F. Tadros, Rheological properties of emulsion systems, in *Emulsions – A*

Fundamental and Practical Approach, ed. J. Sjoblom, NATO ASI Series, Vol. 363, Kluwer, Dordrecht (1991).

21 Y. Saiki and C.A. Prestidge, *Korea–Aust. Rheol. J.*, *17*, 19 (2005).

22 Th.F. Tadros, *Colloids Surf. A*, *91*, 215 (1994).

23 R. Pons, P. Rossi and Th.F. Tadros, *J. Phys. Chem.*, 99, 12624 (1995).

24 Th.F. Tadros, J. Cutler, R. Pons and P. Rossi, Investigations of the interactions between suspensions and emulsions (suspoemulsions), in *Modern Aspects of Colloidal Dispersions*, ed. R.H. Ottewill and A.R. Rennie, Kluwer, Dordrecht (1998).

25 H.M. Princen and A.D. Kiss, *J. Colloid Interface Sci.*, *112*, 427 (1986).

8
Rheology Modifiers, Thickeners and Gels

8.1
Introduction

In any paint formulation, one needs to modify the rheology of the system to achieve the following properties: (i) long-term physical stability (absence of creaming or sedimentation and separation) and (ii) ease of application, good coating and lack of sag.

8.2
Classification of Thickeners and Gels

Rheology modifiers can be classified into several categories:

1. Gels produced as a result of repulsive interaction, e.g. expanded double layers.
2. Self-structured systems, whereby one induces weak flocculation to produce a 'gel' by the particles or droplets. This requires control of the particle size and shape, volume fraction of the dispersion and depth of the secondary minimum.
3. Thickeners consisting of high molecular weight polymers or finely divided particulate systems that interact in the continuous phase forming a 'three-dimensional' structure.
4. Cross-linked polymers (chemical gels).
5. Self-assembled structures such as associative thickeners.
6. Liquid crystalline structures of the hexagonal, cubic or lamellar phases.

8.3
Definition of a 'Gel'

In all the above systems, one essentially produces a 'gel'. A gel is a 'semi-solid' consisting of a 'network' in which the solvent is 'entrapped'. It may be classified as a 'liquid-in-solid' dispersion. A gel shows some solid-like properties and also liquid-like properties, i.e. it is a viscoelastic system. Depending on the gel

Colloids and Interface Science Series, Vol. 6
Colloids in Paints. Tharwat F. Tadros

ISBN: 978-3-527-31466-9

strength, the system may behave as a viscoelastic solid or viscoelastic liquid depending on the stress applied on the gel. For 'strong' gels (such as those produced by chemical cross-linking), the system may behave as a viscoelastic solid up to high stresses and the gel could also show a significant yield value. For 'weaker' gels, e.g. those produced by associative thickeners, the system may show viscoelastic liquid-like behavior at lower applied stresses compared with chemical gels.

8.4
Rheological Behavior of a 'Gel'

8.4.1
Stress Relaxation (After Sudden Application of Strain)

One of the most useful ways to describe a gel is to consider the relaxation time of the system [1]. Consider a 'gel' with the components in some sort of a 'three-dimensional' structure. To deform it instantly, a stress is required and energy is stored in the system (high-energy structure). To maintain the new shape (constant deformation), the stress required becomes smaller since the components of the 'gel' undergo some diffusion, resulting in a lower energy structure to be approached (structural or stress relaxation). At long times, deformation becomes permanent with complete relaxation of the structure (new low-energy structure) and viscous flow will occur.

The above behavior is represented schematically in Figure 8.1, where the stress (after sudden application of strain) is plotted as a function of time. This represen-

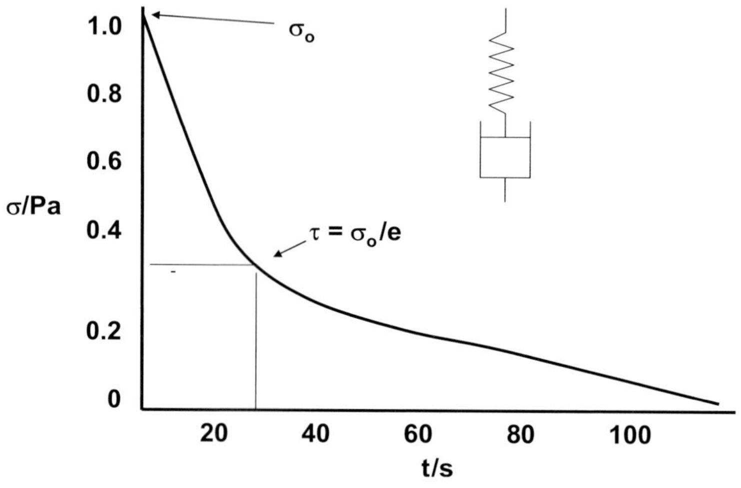

Figure 8.1 Stress relaxation after sudden application of strain for a viscoelastic liquid.

tation is for a viscoelastic liquid (Maxwell element represented by a spring and dash-pot in series) with complete relaxation of the springs at infinite time (see Chapter 6). In other words, the stress approaches zero at infinite time.

The above exponential decay of the stress can be represented by the following equation:

$$\sigma(t) = \sigma_0 \exp\left(-\frac{t}{\tau}\right) \tag{1}$$

where τ is the stress relaxation time.

If the stress is divided by the strain, one obtains the modulus G:

$$G(t) = G_0 \exp\left(\frac{t}{\tau}\right) \tag{2}$$

where G_0 is the instantaneous modulus (the spring constant).

Many cross-linked gels behave like viscoelastic solids (Kelvin model) with another spring in parallel having an elasticity G_e. The modulus does not decay to zero.

The relaxation modulus is given by

$$G(t) = G_0 \exp\left(\frac{t}{\tau}\right) + G_e \tag{3}$$

Figure 8.2 shows the variation of $G(t)$ with time for a viscoelastic solid.

A useful way to distinguish between the various gels is to consider the Deborah number, D_e:

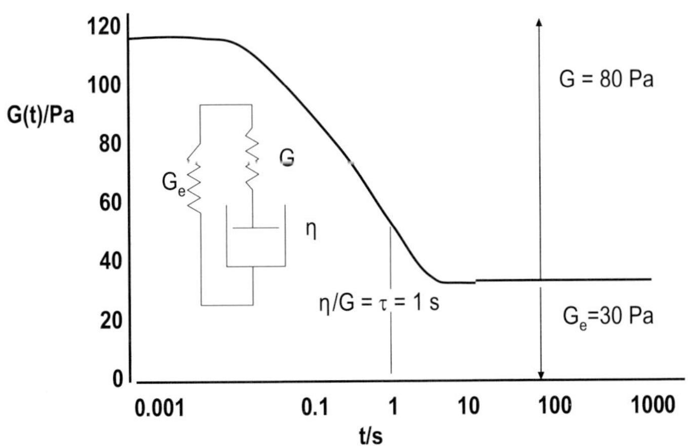

Figure 8.2 Variation of $G(t)$ with time for a viscoelastic solid.

$$D_e = \frac{\tau}{t_e} \tag{4}$$

For a gel that shows 'solid-like' behavior ('three-dimensional structure'), D_e is large compared with a gel that behaves as a viscoelastic liquid.

8.4.2
Constant Stress (Creep) Measurements [1]

In this case, a constant stress σ is applied and the strain (deformation) γ or compliance J ($= \gamma/\sigma$, Pa^{-1}) is followed as a function of time. A gel that consists of a strong 'three-dimensional' structure (e.g. cross-linked) behaves as a viscoelastic solid, as illustrated in Figure 8.3. This behavior may occur up to high applied stresses. In other words, the critical stress above which significant deformation occurs can be fairly high. A weaker gel (produced, for example, by high molecular weight polymers that are physically attached) behaves as a viscoelastic liquid as shown in Figure 8.3. In this case, viscoelastic solid behavior occurs only at much lower stresses than those observed with the cross-linked gels.

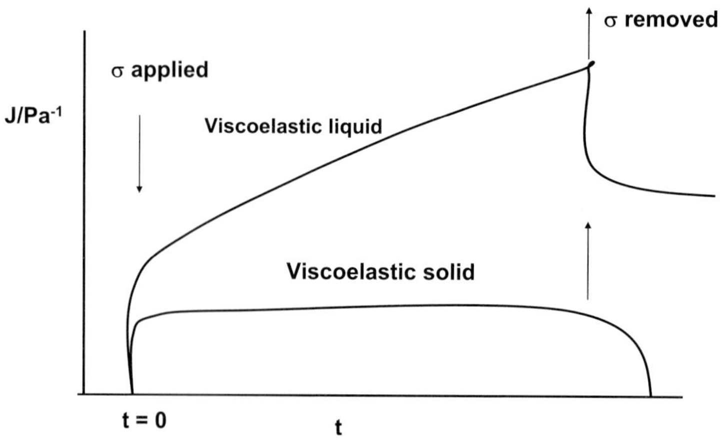

Figure 8.3 Viscoelastic solid and viscoelastic liquid response for gels.

8.4.3
Dynamic (Oscillatory) Measurements [1]

A sinusoidal strain (or stress) with amplitude γ_0 and frequency ω (rad s^{-1}) is applied to the system and the resulting stress (or strain) with amplitude σ_0 is simultaneously measured. This is illustrated in Figure 8.4.

For any gel $\delta < 90°$, and the smaller the value of δ the stronger is the gel. From the amplitudes of stress and strain (σ_0 and γ_0) and the phase angle shift δ, one can obtain the following viscoelastic parameters:

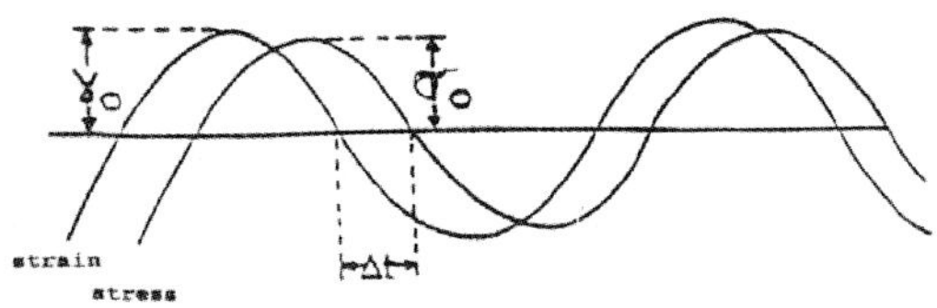

Δ t = time shift for sine waves of stress and strain.

Δ tω = δ phase angle shift

ω = frequency in radian s^{-1}

ω = 2πυ

Perfectly elastic solid	δ = 0
Perfectly viscos liquid	δ = 90°
Viscoelastic system	0 < δ < 90°

Figure 8.4 Sine waves for strain and stress.

$$|G^*| = \frac{\sigma_0}{\gamma_0} \tag{5}$$

$$\text{Storage (elastic) modulus } G' = |G^*| \cos \delta \tag{6}$$

$$\text{Loss (viscous) modulus } G'' = |G^*| \sin \delta \tag{7}$$

$$\tan \delta = \frac{G''}{G'} \tag{8}$$

For gels, $\tan \delta < 1$ and the smaller the value the stronger is the gel.

8.5 Classification of Gels

Gels may be conveniently classified into two main categories:

1. gels based on macromolecules (polymer gels)
2. gels based on solid particulate materials.

Numerous examples of gels based on polymers may be identified: gels produced by overlap or 'entanglement' of polymer chains (physical gels), gels produced by association of polymer chains (the so-called associative thickeners), gels produced by physical or chemical cross-linking of polymer chains (sometimes referred to as 'microgels'), etc. The most common particulate gels are those based on 'swelling' clays (both aqueous and non-aqueous) and finely divided oxides (e.g. silica gels).

In addition to the above two main classes, gels can also be produced from surfactant liquid crystalline phases: hexagonal, cubic or lamellar structures. These gels may be produced from single surfactant molecules, usually at high concentrations (>30%). They can also be produced using mixtures of surfactants and other amphiphiles such as long-chain alcohols, e.g. mixtures of alcohol ethoxylates with cetyl, stearyl or cetostearyl alcohol. Gels can also be produced from ionic surfactants by addition of other ingredients, e.g. salts and/or long-chain alcohols.

8.6 Polymer Gels

8.6.1 Physical Gels Obtained by Chain Overlap

Flexible polymers that produce random coils in solution can produce 'gels' at a critical concentration C^*, referred to as the polymer coil 'overlap' concentration [2]. This picture can be realized if one considers the coil dimensions in solution: considering the polymer chain to be represented by a random walk in three dimensions, one may define two main parameters, namely the root mean square end-to-end length $\langle r^2 \rangle^{\frac{1}{2}}$ and the root mean square radius of gyration $\langle s^2 \rangle^{\frac{1}{2}}$ (sometimes denoted R_G). The two are related by

$$\langle r^2 \rangle^{\frac{1}{2}} = 6^{\frac{1}{2}} \langle s^2 \rangle^{\frac{1}{2}} \tag{9}$$

The viscosity of a polymer solution increases gradually with increase in its concentration and at a critical concentration, C^*, the polymer coils with a radius of gyration R_G and a hydrodynamic radius R_h (which is higher than R_G due to solvation of the polymer chains) begin to overlap, and this shows a rapid increase in viscosity. This is illustrated in Figure 8.5, which shows the variation of log η with log C.

In the first part of the curve, $\eta \propto C$, whereas in the second part (above C^*), $\eta \propto C^{3.4}$. A schematic representation of polymer coil overlap is presented in Figure 8.6, which shows the effect of gradually increasing the polymer concentration. The polymer concentration above C^* is referred to as the semi-dilute range [2].

C^* is related to R_G and the polymer molecular weight M by

$$C^* = \left(\frac{4}{3}\right) \pi {R_G}^3 \left(\frac{N_{av}}{M}\right) \approx 1 \tag{10}$$

where N_{av} is the Avogadro's number. As M increases, C^* becomes progressively lower. This shows that to produce physical gels at low concentrations by simple polymer coil overlap, one has to use high molecular weight polymers.

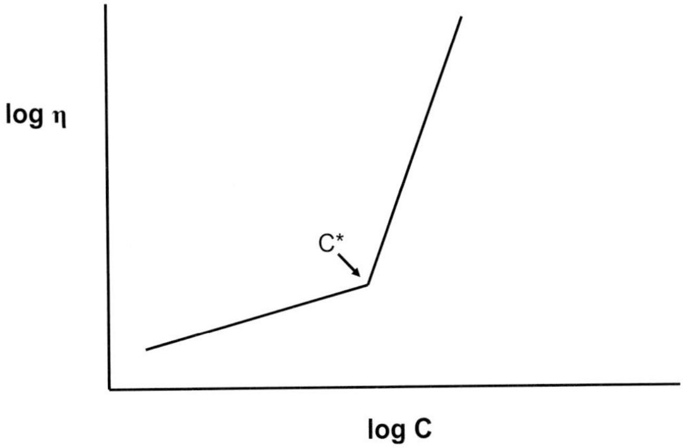

Figure 8.5 Variation of log η with log C.

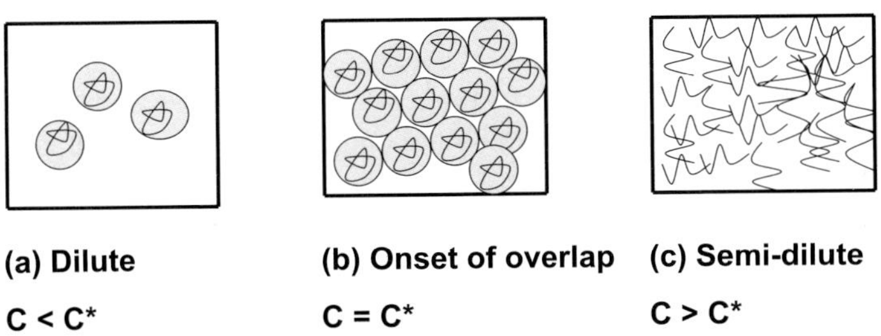

Figure 8.6 Cross-over between dilute and semi-dilute solutions.

Another method to reduce the polymer concentration at which chain overlap occurs is to use polymers that form extended chains such as xanthan gum, which produces conformation in the form of a helical structure with a large axial ratio. These polymers give much higher intrinsic viscosities and they show both rotational and translational diffusion. The relaxation time for the polymer chain is much higher than that for a corresponding polymer with the same molecular weight, but produces a random coil conformation.

The above polymers interact at very low concentrations and the overlap concentration can be very low (<0.01%). These polysaccharides are used in many formulations to produce physical gels at very low concentrations.

8.6.2 Gels Produced by Associative Thickeners

Associative thickeners are hydrophobically modified polymer molecules where alkyl chains (C_{12}–C_{16}) are either randomly grafted on a hydrophilic polymer molecule such as hydroxyethylcellulose (HEC) or simply grafted at both ends of the hydrophilic chain. An example of hydrophobically modified HEC is Natrosol Plus (Hercules), which contains 3–4 C_{16} units randomly grafted on to HEC. Another example of a polymer that contains two alkyl chains at both ends of the molecule is HEUR (Rohm and Haas), which is made of poly(ethylene oxide) (PEO) that is capped at both ends with a linear C_{18} hydrocarbon chain.

The above hydrophobically modified polymers form gels when dissolved in water. Gel formation can occur at relatively lower polymer concentrations compared with the unmodified molecule.

The most likely explanation of gel formation is that it is due to hydrophobic bonding (association) between the alkyl chains in the molecule. This effectively causes an apparent increase in the molecular weight. These associative structures are similar to micelles, except the aggregation numbers are much smaller.

Figure 8.7 shows the variation of viscosity (measured using a Brookfield viscometer at 30 rpm) as a function of the alkyl content (C_8, C_{12} and C_{16}) for hydrophobically modified HEC (HMHEC). The viscosity reaches a maximum at a given alkyl group content that decreases with increase in the alkyl chain length. The viscosity maximum increases with increase in the alkyl chain length.

Associative thickeners also show interaction with surfactant micelles that are present in the formulation. The viscosity of the associative thickeners shows a maximum at a given surfactant concentration that depends on the nature of surfactant. This is shown schematically in Figure 8.8.

The increase in viscosity is attributed to the hydrophobic interaction between the alkyl chains on the backbone of the polymer with the surfactant micelles. A

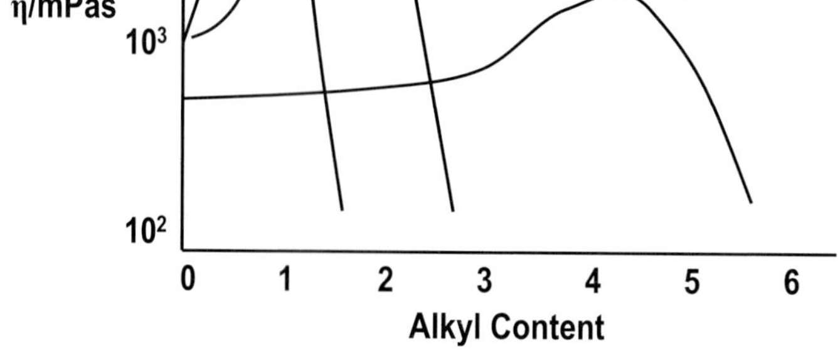

Figure 8.7 Variation of viscosity of 1% HMHEC versus alkyl group content of the polymer.

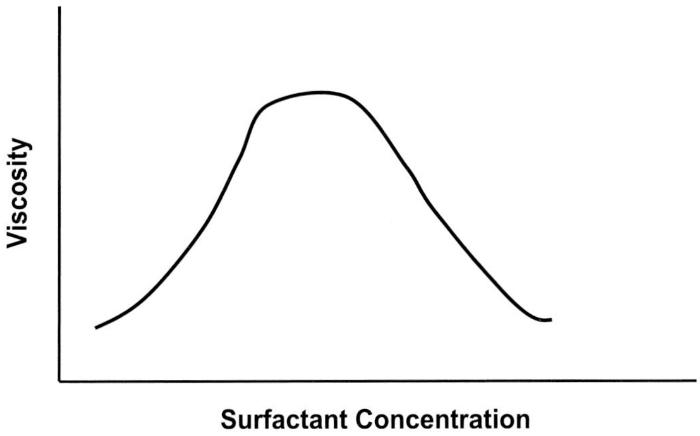

Figure 8.8 Schematic plot of viscosity of HM polymer versus surfactant concentration.

schematic diagram showing the interaction between HM polymers and surfactant micelles is shown in Figure 8.9.

At higher surfactant concentration, the 'bridges' between the HM polymer molecules and the micelles are broken (free micelles) and η decreases [3].

The viscosity of HM polymers shows a rapid increase at a critical concentration, which may be defined as the critical aggregation concentration (CAC) as illustrated in Figure 8.10 for HMHEC (WSP-D45 from Hercules). The assumption is made that the CAC is equal to the coil overlap concentration C^*.

From a knowledge of C^* and the intrinsic viscosity $[\eta]$, one can obtain the number of chains in each aggregate. For the above example, $[\eta] = 4.7$ and $C^*[\eta] = 1$, giving an aggregation number of $\sim$4

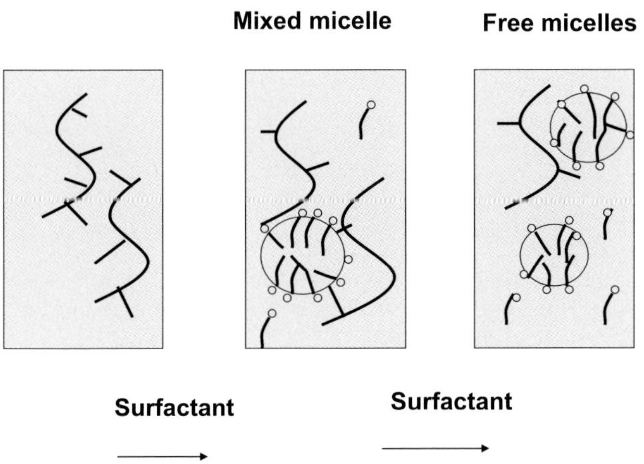

Figure 8.9 Schematic representation of the interaction of polymers with surfactants.

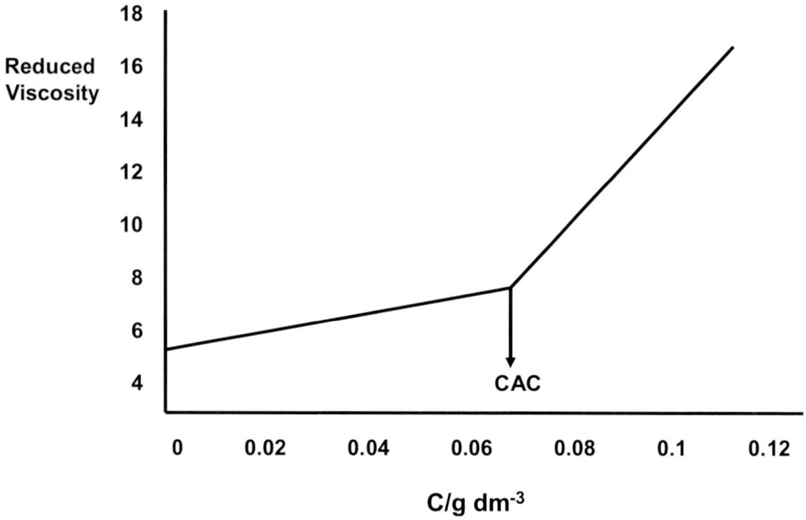

Figure 8.10 Variation of reduced viscosity with HMHEC concentration.

At C^*, the polymer solution shows non-Newtonian flow (shear thinning behavior) and it shows a high viscosity at low shear rates. This is illustrated in Figure 8.11, which shows the variation of apparent viscosity with shear rate (using a constant stress rheometer). Below ~0.1 s^{-1}, a plateau viscosity value $\eta(0)$ (referred to as residual or zero shear viscosity) is reached (~200 Pa s).

With increase in polymer concentration above C^*, the zero shear viscosity increases with increase in polymer concentration. This is illustrated in Figure 8.12.

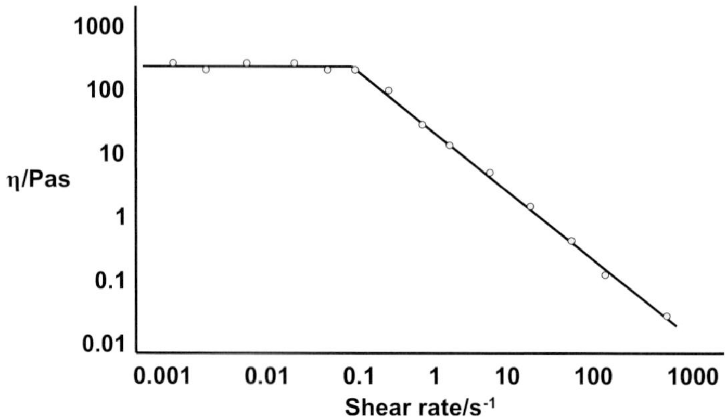

Figure 8.11 Variation of viscosity with shear rate for HMHEC WSP-47 at 0.75 g per 100 cm^{-3}.

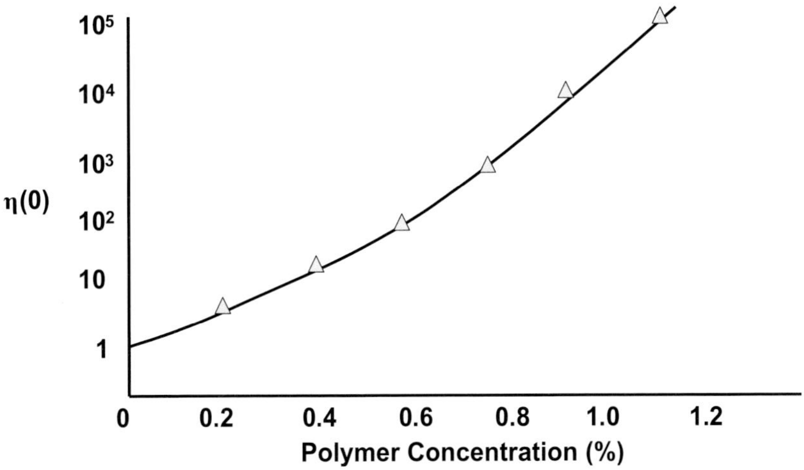

Figure 8.12 Variation of $\eta(0)$ with polymer concentration.

The above hydrophobically modified polymers are viscoelastic. This is illustrated in Figure 8.13 for a solution 5.25% of C_{18} end-capped PEO with $M =$ 35 000, which shows the variation of the storage modulus G' and loss modulus G'' with frequency ω (rad s^{-1}). G' increases with increase in frequency and ultimately it reaches a plateau value at high frequency. G'' (which is higher than G' in the low-frequency regime) increases with increase in frequency, reaches a maximum at a characteristic frequency ω^* (at which $G' = G''$) and then decreases to near zero in the high-frequency regime. This variation of G' and G'' with ω is typical for a system that shows Maxwell behavior.

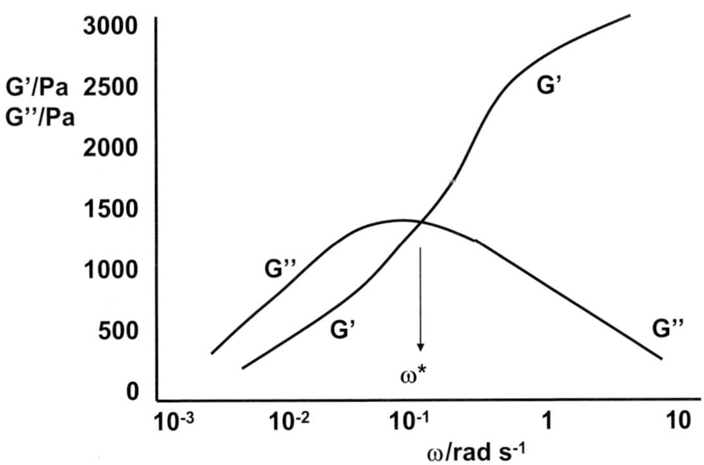

Figure 8.13 Variation of G' and G'' with frequency for 5.25% HM PEO.

From the cross-over point ω^* (at which $G' = G''$), one can obtain the relaxation time τ of the polymer in solution:

$$\tau = \frac{1}{\omega^*} \tag{11}$$

For the above polymer, $\tau = 8$ s.

The above gels (sometimes referred to as rheology modifiers) are used in many formulations to produce the right consistency and also for reduction of sedimentation or creaming of suspensions and emulsions. These hydrophobically modified polymers can also interact with hydrophobic particles in a suspension, forming several other associative structures.

The high-frequency modulus, sometimes referred to as the network modulus, can be used to obtain the number of 'links' in the gel network structure. Using the theory of rubber elasticity, the network modulus G_N is related to the number of elastically effective links N and a factor A that depends on the junction functionality:

$$G_N = ANkT \tag{12}$$

where k is the Boltzmann constant and T is the absolute temperature.

For an end-capped PEO (e.g. HEUR), the junctions should be multifunctional ($A = 1$); for tetra-functional junctions $A = \frac{1}{2}$.

8.6.3 Cross-linked Gels (Chemical Gels)

Many commercially available gels are made using cross-linking agents to produce what are sometimes referred to as 'microgels'. These microgel particles are dispersed in the liquid and they undergo solvent swelling, which may also be enhanced by some chemical modification, e.g. pH adjustment in aqueous systems.

A typical example of cross-linked gels is that based on poly(acrylic acid), which is commercially available under the trade name Carbopol (B.F. Goodrich). These microgel particles are dispersed in aqueous solution and, on neutralization with NaOH or ethanolamine, they swell as a result of the ionization of the poly(acrylic acid) chains. The ionization of poly(acrylic acid) occurs when the pH is increased above ~5 and these ionized chains form extended double layers, forming gels at low microgel concentration (mostly less than 1% of the microgel particles).

Another example of microgels is that based on *N*-isopropylacrylamide cross-linked with *N*,*N*′-methylenebisacrylamide (poly-NIPAM). These microgel particles are swollen by temperature changes. At temperatures > 35 °C, the cross-linked polymer is in a collapsed state. When the temperature is reduced, the cross-linked polymer swells, absorbing water that increases its volume by several order of magnitudes. These polymer gels are sometimes referred to as 'smart' colloids and they have several applications in controlled release.

8.7 Particulate Gels

Two main interactions can cause gel formation with particulate materials:

- *long-range repulsions between the particles*, e.g. using extended electrical double layers or steric repulsion resulting from the presence of adsorbed or grafted surfactant or polymer chains
- *van der Waals attraction between the particles* (flocculation), which can produce three-dimensional gel networks in the continuous phase.

All of the above systems produce non-Newtonian systems that shows a 'yield value' and high viscosity at low shear stresses or shear rates.

Some examples may be given to illustrate the above particulate gels:

1. *Swellable clays*, e.g. sodium montmorillonite (sometimes referred to as bentonite) at low electrolyte concentration [4]. These produce gels as a result of the formation of extended double layers. At moderate electrolyte concentrations, the clay particles may form association structures as a result of face-to-edge flocculation (see below). The above clays can be modified by interaction with alkylammonium salts (cationic surfactants) to produce hydrophobically modified clays sometimes referred to as organo-clays or bentones. These can be dispersed in non-aqueous media and swollen by addition of polar solvents.
2. *Finely divided oxide*, e.g. silica, which can produce gels by aggregation of the particles to form three-dimensional gel structures.

In many cases, particulate solids are combined with high molecular weight polymers to enhance gel formation, e.g. as a result of 'bridging' or 'depletion flocculation'.

8.7.1 Aqueous Clay Gels

Gel formation using swellable clays such as sodium montmorillonite can be understood from the structure of the clay mineral and interparticle interaction in aqueous solutions [4]. These clay minerals are formed from very thin plates (1 nm thickness) formed from two layers of tetrahedral silica with one octahedral alumina sheet (between the two silica layers). The charge in clay mineral is produced by a process referred to as isomorphic substitution, whereby atoms with high valency are replaced with ions of lower valency (e.g. an S^{4+} ion is replaced by an Al^{3+} ion). This produces a negative charge on the surface of the platelet that is compensated by Na^+ ions. The edges of the clay platelets contains oxide-like material, e.g. Al–OH, which can acquire a positive charge at pH < 7 (the iso-electric point of Al–OH is pH ≈ 7–9).

Various interactions between clay particles in aqueous solution can be produced depending on the pH and electrolyte concentration. At very low electrolyte concentration, the double layers are extended and gels are produced when these

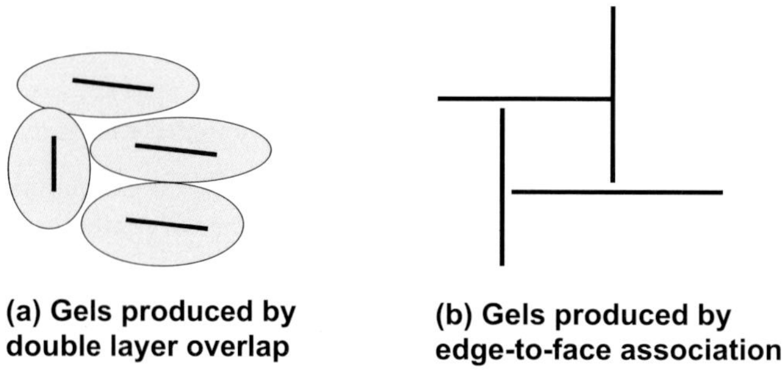

Figure 8.14 Schematic representation of gel formation in aqueous clay dispersions.

double layers begin to overlap. At intermediate electrolyte concentrations, the double layers at the faces and edges are compressed and interaction may take place by edge-to-face association (making T-junctions, referred to as a 'house of cards' structure). This also produces a gel in aqueous solution.

The above two types of interactions are shown schematically in Figure 8.14.

Evidence for the above structures may be obtained using rheological measurements. Figure 8.15 shows the variation of yield value σ_β (for a 3.22% clay dispersion) as a function of electrolyte concentration. When $C = 0$, the double layers are extended and gel formation is due to double layer overlap (Figure 8.14a). First addition of NaCl causes compression of the double layers and hence the yield value decreases very rapidly. At intermediate NaCl concentrations, gel formation occurs as a result of face-to-edge association (house of cards structure) and the yield value increases very rapidly with increase in NaCl concentration. If the NaCl concentration is increased further, face-to-face association may occur and the yield value decreases (the gel is destroyed).

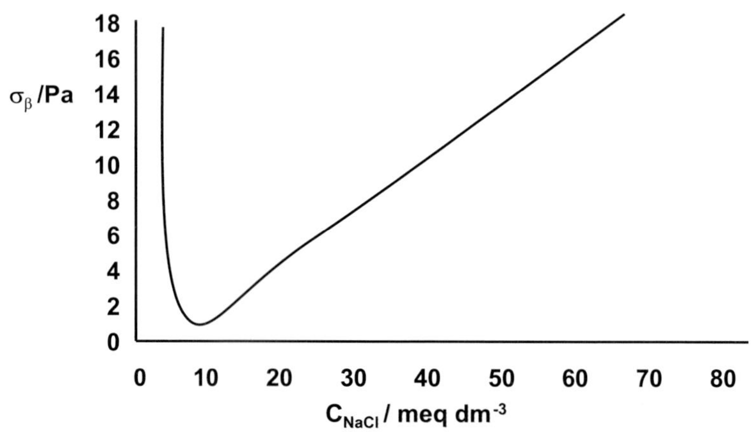

Figure 8.15 Variation of yield value with NaCl concentration for 3.22% sodium montmorillonite dispersions.

8.7.2
Organo-clays (Bentones)

These are produced by exchanging the Na^+ ions with alkylammonium ions, e.g. dodecyl- or cetyltrimethylammonium ions. In some cases, dialkylammonium ions are also used. In this case, the clay particle surface will be covered with hydrophobic alkyl groups and hence it can be dispersed in organic solvents, e.g. hydrocarbon or silicone oils. The exchange is not carried out completely, leaving few hydrophilic groups on the surface. The dispersed organo-clays are then activated by addition of a polar solvent such as propylene carbonate, alcohols or glycols.

The gel is produced by hydrogen bonding between the polar groups on the surface of the clay and the polar solvent added. Several types of organo-clays are commercially available depending on the application and the type of solvent in which a gel is required. In some cases, organo-clays that are already activated can be supplied.

Organo-clays are applied to 'thicken' many personal care products, e.g. foundations, non-aqueous creams, nail polish and lipsticks. The procedure for dispersion of the organo-clay particles and their subsequent activation is crucial and required good process control.

8.7.3
Oxide Gels

The most commonly used oxide gels are based on silica. Various forms of silicas can be produced, the most common being referred to as fumed and precipitated silicas. Fumed silicas such as Aerosil 200 are produced by reaction of silicon tetrachloride with steam. The surface contains siloxane bonds and isolated silanol groups (referred to a vicinal). Precipitated silicas are produced from sodium silicate by acidification. The surface is more populated with silanol groups than fumed silica. It contains geminal OH groups (two attached to the same Si atom). Both fumed and precipitated silicas can produce gels in both aqueous and non-aqueous systems. Gelation results from aggregation of silica particles, thus producing three-dimensional gel networks with a yield value.

In aqueous media, the gel strength depends on the pH and electrolyte concentration [5]. As an illustration, Figure 8.16 shows the variation of viscosity and yield value with concentration of Aerosil silica (which has been dispersed by sonication) at three different pH values. In all cases, the viscosity and yield value shows a rapid increase above a certain silica concentration that depends on the pH of the system.

At pH 3 (near the isoelectric point of silica), the particles are aggregated (forming flocs) and the increase in viscosity occurs at relatively high silica concentration (>11% v/v). At pH 7, the silica becomes negatively charged and the double layers stabilize the silica particles against aggregation. In this case, the particles remain as small units and the viscosity and yield value increase sharply above 7% v/v. At pH 9, some aggregation occurs as a result of the electrolyte released

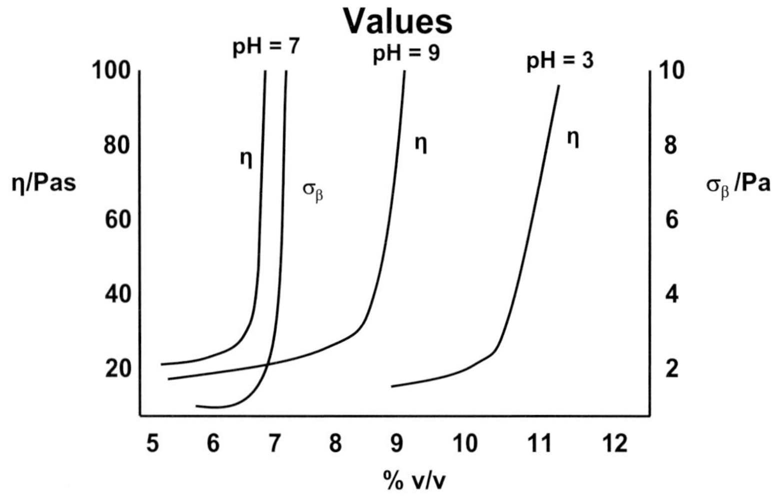

Figure 8.16 Variation of viscosity η and yield value σ_β with Aerosil 200 concentration at pH 3, 7 and 9.

on adjusting the pH; in this case, the viscosity increases at higher concentration (>9% v/v) compared with the results at pH 7.

The above results clearly indicate the importance of pH and electrolyte concentration in the gelation of silica. It seems that the optimum gel formation occurs at neutral pH.

Silica gels are used in many personal care applications, in particular for control of the rheology of toothpaste. Hydrophilic silica (Aerosil 200) can also be applied for forming gels in non-aqueous media. In this case, gel formation is the result of hydrogen bond formation between the particles which produce three-dimensional structures in the non-aqueous medium.

8.7.4
Gels Produced Using Particulate Solids and High Molecular Weight Polymers

In many cases, particulate materials are combined with high molecular weight polymers to produce three-dimensional structures by association of the polymer with the particles. Several mechanisms have been suggested for gel formation in these mixtures, e.g. bridging by the polymer chains and depletion flocculation.

The above mixtures give more robust gel structures which in many case have a lower temperature dependence compared with the individual components. The optimum composition of these particulate–polymer mixtures can be obtained using rheological measurements [6].

By measuring the yield value as a function of polymer concentration at a fixed particulate concentration, one can obtain the optimum polymer concentration required. In most cases, the yield value reaches a maximum at a given ratio of particulate solid to polymer. This trend may be due to bridging flocculation, which reaches an optimum at a given surface coverage of the particles (usually at 0.25–0.5 surface coverage).

All the above-mentioned gels produce thixotropy, i.e. a reversible decrease in viscosity on application of shear (at a constant shear rate) and recovery of the viscosity on standing. This thixotropic behavior finds application in many systems in personal care products, e.g. in creams, toothpastes and foundations.

One of the most useful techniques to study thixotropy is to follow the change in modulus with time after application of shear, i.e. after subjecting the dispersion to a constant shear rate, oscillatory measurements are carried out at low strains and high frequency and the increase in modulus with time (which is exponential) can be used to characterize the recovery of the gel.

References

1 J.D. Ferry, *Viscoelastic Properties of Polymers*, John Wiley & Sons, Inc., New York (1980).
2 P.G. de Gennes, *Scaling Concepts in Polymer Physics*, Cornell University Press, Ithaca, NY (1979).
3 E.D. Goddard, in *Polymer/Surfactant Interaction*, ed. E.D. Goddard and J.V. Gruber, Marcel Dekker, New York, Chapters 4 and 5 (1999).
4 H. Van Olphen, Clay Colloid Chemistry, John Wiley & Sons, New York (1961).
5 D. Heath and Th.F. Tadros, *J. Colloid Interface Sci.*, 93, 307, 320 (1983).
6 Th.F. Tadros, *Applied Surfactants*, Wiley-VCH Verlag GmbH, Weinheim (2005).

9
Rheology of Paints

9.1
Introduction

As mentioned in Chapter 1, paints are complex colloidal dispersions of solid and 'liquid' (latex) particles that are dispersed in a liquid medium (continuous phase), which could be aqueous or non-aqueous depending on the applications. The continuous phase also contains various rheological modifiers, which could be polymers or inert, fine, solid particles (see Chapter 8). The interaction between the various components in a paint formulation results in a non-Newtonian system with complex rheological behavior. Control of paint rheology is essential for successful utilization of the paint. Whatever application technique is used, e.g. spray gun, brush, roller, three stages must be considered when considering the rheology of a paint [1]: (i) transfer of the paint from the bulk container to the applicator; (ii) transfer of the paint to the surface to form a thin and even film, i.e. film formation; and (iii) flow-out of the film surface, coalescence of the polymer latex particles and loss of the medium by evaporation. Each of these processes requires accurate control of the rheological characteristics.

In the bulk container, the paint should be of sufficiently low viscosity that it can be readily utilized in the applicator. For application by a brush or a hand roller, the paint should readily penetrate the spaces between the bristles of a brush or the porous surface of the roller. The paint is then held by capillary/surface tension forces during the transfer to the surface to be painted. Control of the brush loading is crucial in any paint application. If the brush loading is too high, the total weight of the paint becomes sufficient to overcome the capillary forces, leading to paint drip or run-off from the brush, clearly an undesirable result. In contrast, if the brush load is too low, it results in a thin paint film, or a non-uniform film with thicker film over a smaller surface area [1]. To achieve the optimum film thickness, one should control the flow-out properties, as will be discussed below.

Increased brush loading may be achieved by increasing the bulk paint viscosity or by introducing rheology modifiers (see Chapter 8). The latter is the preferred option, since these rheology modifiers produce a shear thinning system, whereby the viscosity of the paint is rapidly reduced during application, thus reducing the

Colloids and Interface Science Series, Vol. 6
Colloids in Paints. Tharwat F. Tadros

ISBN: 978-3-527-31466-9

mechanical effort required to spread into a film. The recovery of the viscosity after application prevents dripping or running of the paint. In addition, these rheology modifiers produce high residual viscosity, thus preventing sedimentation of particles in the paint.

In industrial applications such as spraying or roller coating, control of the paint rheology is crucial. In spraying, the viscosity of the bulk paint must be low enough to allow the paint to be pumped through the fine jet of the spray gun with minimal application of pressure. In most cases, the paint is thinned from the higher solids bulk immediately before application, and during such a short period, settling of the particles is less of a problem. The rheology of the diluted paint has a large influence on the droplet spectrum of the sprayed paint [1]. One should avoid the formation of small spray droplets which may undergo drift on application. The use of rheological modifiers is essential to produce the optimum droplet size distribution in the spray.

In the roller-coating process, the paint must be considerably thicker than that used in spray processes. The paint should be able to flow under gravity or low pumping energy to the surface of the application roller, where it may be spread to an even layer by the action of a doctor blade or another roller. In this case, the mechanical work required to cause the paint to flow is much less important. However, the paint must be viscous enough to prevent it running off or being thrown off the roller by centrifugal force.

In both spraying and roller-coating applications, the fluid flow rate and operating speed are very high and in this case both high stresses and high shear rates operate in the process. One should also note that the paint remains in the spray gun or the 'nip' between the rollers for a very short time, hence a steady state is never reached. In this case only transient or high-frequency rheological measurements (see Chapter 6) are likely to produce relevant rheological parameters. Shear rates as high as 10^5 s^{-1} can be reached in high-speed rollers. At such high shear rates and in the presence of high molecular weight polymers in the paint formulation, a high extensional viscosity (which can be several orders of magnitude higher than the shear viscosity) can develop. The extensional viscosity of 'thickened' water-based emulsion paints influences the application properties such as tracking and spattering. One would expect that such a high extensional viscosity can interfere with the process of filament or jet rupture to form spray droplets [2].

The next process that must be considered in paint application is film formation. The loading and transfer of the paint by a brush or a roller from the bulk container to the surface to be painted are followed by regular movement of the brush or roller over the surface to transfer the load of paint from the brush to the surface and spread it out in an even layer. During this process, hand pressure on the brush causes shearing and compression of the brush bristles or fibers of the rubber foam or fibrous mat typically covering the surface of a hand roller. The flow processes involved are very complex and very difficult to analyze. However, some attempts have been made to calculate the range of shear rates involved in paint brush applications. Ranges of 15–30 s^{-1} were estimated for brush dipping and 2500–10 000 s^{-1} for brush spreading [3, 4].

The third and most important step of paint application is that of flow-out or leveling of the paint film, which involves latex coalescence and loss of medium by evaporation. This has a major influence on color uniformity, hiding power and major flow faults such as sagging and slumping [5–7]. Unfortunately, there is still a lack of understanding of the relevant theological parameters that affect these processes. There may be some correlation between elastic recovery and surface irregularity flow-out (leveling). Both pigment dispersions (mill bases) and the final paint formulation show viscoelastic behavior [8, 9]. This clearly demonstrates the importance of dynamic (oscillatory) measurements in the assessment of leveling properties.

Solvent evaporation during application will have a major effect on the rheological properties and also the surface tension at the wet film/air interface. Evaporation results in an increase in the disperse volume fraction and a cooling effect at the film surface. Both effects lead to a tangential surface shearing force. It has been argued that the hydrostatic pressure gradient in a paint film is insufficient to explain the leveling effect. Whereas surface tension tends to produce a flat surface, irrespective of the substrate surface profile underneath, the surface tension gradient developing over the wet film tends to produce a uniform film thickness, i.e. the surface profile of the paint film mirrors exactly the surface profile of the substrate.

Solvent evaporation also leads to gradients in the solvent concentration through the film and across the surface. This leads to density gradients which, with the surface tension gradients, could contribute to circular patterns being set up in the wet paint film. This may lead to the formation of Bernard cell patterns commonly observed at the surface of boiling or rapidly evaporating bulk liquid samples.

The complexity of the rheology of the applied paint film is shown by its viscoelastic behavior and the non-linear and time-dependent effects arising from the high shear during application. In addition, due to the concentration gradients through the film thickness, the rheology will also vary through the depth of the film. Rough calculations showed that the operative forces in leveling are in the range 3–5 Pa and in sagging about 0.8 Pa at the surface of a typical paint film. The shear rates for leveling processes in paint films are in the range 10^{-3}–5×10^{-1} s^{-1}. Since the shear stress resulting from gravitational and surface tension forces control the flow in leveling and sagging, it is important to carry out constant stress (creep) measurements when considering the relation between paint rheology and its flow characteristics. Measurement of shear rates seems to be irrelevant in this case.

From the above discussion, one can summarize the desirable rheological characteristics of a paint formulation. It should have a sufficiently low viscosity to facilitate its transfer to the applicator. However, it should also have a sufficiently high residual (zero shear) viscosity to prevent particle sedimentation and a sufficiently high modulus to prevent separation (syneresis). This requires the use of an appropriate rheology modifier (see Chapter 8), which produces a 'gel' structure that is easily 'broken' during transfer from the bulk container. Because of the high shear rates and short time scales involved in the transfer process, both elasticity and extensional flow processes may modify the pattern of surface irregularities on the paint film. The paint must remain low in viscosity for a sufficient time for the surface irregularities to flow out to an acceptable extent. However, while the viscosity is low, the paint may flow on vertical surfaces under the influence of gravity. If the film thickness (film depth) builds up too much, sagging may become noticeable to the observer and lumps of thickened paint may result in an irregular film, which is undesirable. To prevent this from occurring, the initial low viscosity must be followed by a sharp rise in viscosity either by solvent evaporation and/or elastic recovery. The drying film becomes immobilized and sagging is prevented.

9.2
Experimental Techniques for Studying Paint Rheology

Several techniques can be applied for studying paint rheology. During the early stages of paint formulation, i.e. during research and development, one must carry out carefully controlled rheological methods as described in Chapter 6. The results from these controlled experiments can be applied to arrive at the optimum composition of a given paint formulation. Once the latter is established, much simpler methods must be used for quality control of the paint during manufacture. This section starts with the simple methods for quality control. This is then followed by a comprehensive list of the rheological methods that can be applied during the research and development stage.

9.2.1
Experimental Methods for Quality Control

These methods must be fast, reliable and convenient to apply during paint manufacture. Three methods can be applied, namely by measuring the flow through constrictions, by measuring the speed of an object moving through the paint or by measuring the relative speed of motion in a finite sample, where the moving object can be shaped in the form of a spindle which is made to rotate in a fixed volume of the paint. Below, a brief description of the three methods is given. This is followed by a section on measurement of rheology during flow-out (leveling).

Flow through constrictions is best illustrated by the Ford cup, which has been extensively used in the paint industry [1]. A known volume of the paint is held within a vertical cylindrical cup, the bottom of which has a short capillary of controlled length and diameter. The paint is released to flow throw the hole in the bottom of the cup (usually by the operator releasing their finger!) and the time for the paint to flow out of the cup is measured with a stop-watch. The flow end point is normally taken as the point at which the continuous liquid jet breaks up into drops. However, this simple technique suffers from several disadvantages. First, because the liquid height varies during the test, the gravity force driving the liquid through the capillary also varies. Since the paint is non-Newtonian, the viscosity results can be misleading. Second, since the capillary is short, stable flow conditions within the capillary are not obtained. This effect, together with the entry and exit errors, may affect the result, particularly if the paint is elastic in nature. Third, the presence of abrasive particles in the paint may lead to wear of the metal capillary. It is necessary, therefore, to check the Ford cup frequently using Newtonian liquids (e.g. silicone oils) of known viscosity.

Measuring the speed of an object moving through the paint is best illustrated by the Hoeppler-type viscometer, whereby a solid ball of varying density serves as the object. The fall of the ball can be electrically timed between two contacts and the time is a measure of the consistency of the material.

Perhaps the most useful (and commonly used) method for studying paint rheology is based on measuring the relative speed of motion of a spindle rotating in a fixed volume of the paint, as illustrated by the Brookfield viscometer (see Chapter 7). Several spindles and rotation speeds are used to cover a wide range of viscosity. However, a limited range of shear rates is possible when using the Brookfield viscometer. An alternative instrument that is widely used in the paint industry is the Stormer viscometer, having a paddle as a rotational member.

9.2.2 Measurement of Film Flow-out (Leveling and Sagging)

This is perhaps the most difficult method for direct measurement of paint film rheology during flow-out (leveling), because the rheology of paint film material is extremely complex, being not only viscoelastic but also extremely non-linear. In addition, the rheology of a paint film can change rapidly with time, due to solvent evaporation, increase in solids volume fraction and rheological structure recovery. The paint film may also become inhomogeneous in composition through the film depth. For these reasons, it is essential to have rapid methods for following the paint film rheology during flow-out. Several methods can be applied and these are summarized below.

9.2.2.1 Impact Method (Bouncing Ball)

In this method, a 0.5 cm diameter steel ball (weight 0.5 g) is dropped on to a 1.25 cm thick glass slab coated with the paint under test [10]. The rebound height is measured as a function of time as the film dries. Initially, the rebound height decreases with time, since the viscosity of the film increases owing to solvent loss. Consequently, the energy dissipated by the film during the impact of the ball with the glass also increases and this results in a sharp decrease in the rebound height in this early stage of drying. However, as the film cures (by either autoxidation or 'lacquer-type' drying), the film develops some elasticity and the rebound height increases. A schematic representation of the rebound height with drying time is given in Figure 9.1. Using, for example, the ball momentum and energy losses, it is not easy to derive a relationship between rebound height and film viscosity. This is due to the additional factors such as the hydrodynamic force (which

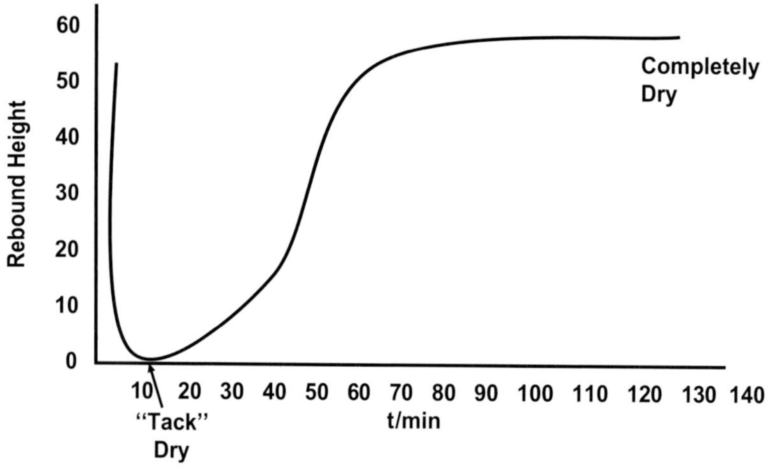

Figure 9.1 Variation of rebound height with time for a 200 μm refinish paint film on glass.

prevents the ball from actually touching the substrate surface) and the paint elasticity effects at the short times of impact (in the region of few tenths of a millisecond). In spite of these drawbacks, the method is simple and one can use paint films on different substrates (glass, metal, wood, etc.) and balls of different size.

9.2.2.2 Impedance Method at High Frequency

The mechanical impedance of an elastic shear wave propagating through a medium is changed by the presence of a viscoelastic layer at the surface of the medium. If the elastic wave is completely damped in this layer, the change in the characteristic impedance can be related to the rheological parameters of the layer material. This method can be applied to follow the changes in the paint film rheology during drying and curing. Pulses of high-frequency oscillations (in the range 2–100 MHz) are generated by means of a suitably-excited piezoelectric crystal attached to the support medium [11]. After propagation through the support, the attenuated pulses are again transformed into electrical signals by a piezoelectric crystal attached to the support. This is shown schematically in Figure 9.2. The

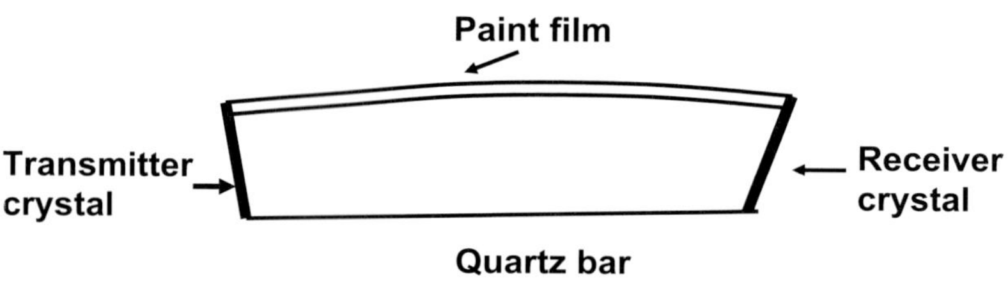

Figure 9.2 Schematic representation of the impedance technique.

phase angle and attenuation of the received pulses are measured and changes in their values are used to compare the rate changes in drying and curing of different paint films. Unfortunately, this technique is limited to measurement of the film properties during only a limited part of the total drying/curing process. In addition, the adhesion of the drying paint film to the substrate can have a major influence on the results.

9.2.3 Rheological Techniques for Research and Development of a Paint System

The basic principles of rheological methods and the techniques that can be applied for their investigation have been described in detail in Chapter 6. In this section, a brief description of these methods that can be applied for a paint formulation is given. Essentially, three main rheological methods can be applied: (i) steady-state shear stress–shear rate measurements with particular attention to time effects (thixotropy); (ii) transient methods: application of constant strain and following the relaxation of the stress with time (stress relaxation measurements) or application of constant stress and following the change of strain with time (creep measurements); and (iii) dynamic (oscillatory) measurements. In addition to the above methods, two main investigations that are relevant to paint systems must be considered: (iv) normal force measurements and (v) elongational viscosity measurements.

9.2.4 Steady-state Shear Stress–Shear Rate Measurements

In these measurements, one applies different, constant rates of shear to the material and measures the resulting stress [12–15]. To calculate the shear stress–shear rate relationship, one should have a well-defined geometry, the most common being the concentric cylinder, the cone and plate and the parallel plate configurations (see Chapter 6). Most paint systems show a pseudoplastic behavior, as illustrated in Figure 9.3, where the stress σ and viscosity η are plotted as a function of shear rate $\dot{\gamma}$.

The latter curve shows the shear thinning behavior of the system. It shows two plateaus (Newtonian) regions at low and high shear rate ranges. The low shear rate plateau region gives the residual (sometimes referred to as 'zero shear rate') viscosity $\eta(0)$ whereas the high shear rate plateau region gives the lowest viscosity that is reached with a shear thinning system, sometimes referred to as $\eta(\infty)$. These two regions are separated by the shear thinning regime, where the viscosity decreases with increase in shear rate. The residual viscosity $\eta(0)$ is an important parameter that determines particle sedimentation. A minimum value is needed to prevent sedimentation. The critical shear rate above which a paint shows the shear thinning behavior is an important parameter that controls the transfer of the paint to the applicator. This critical shear rate should not be too high otherwise the transfer of the paint from the container (which requires a

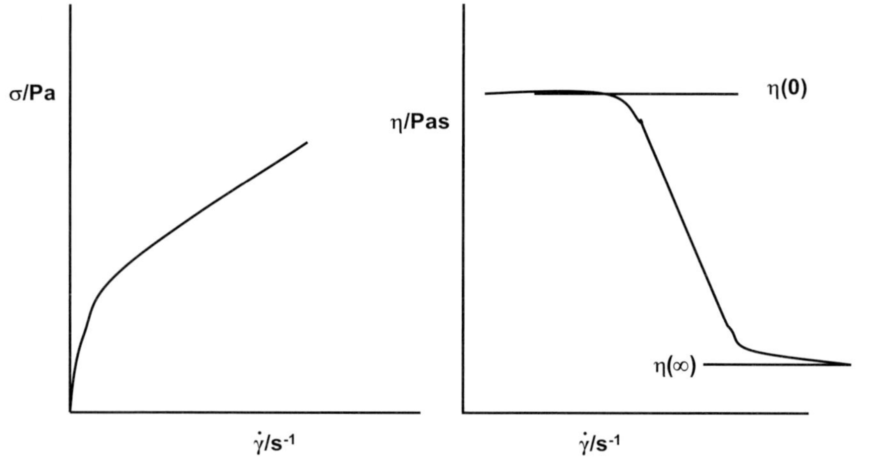

Figure 9.3 Stress–shear rate and viscosity–shear rate relationships for a pseudoplastic system.

low viscosity) becomes very difficult. The high shear rate viscosity $\eta(\infty)$ determines the flow of the paint on the substrate.

Several models have been suggested for analyzing the flow curves of pseudoplastic systems (see Chapter 6) and these are summarized below.

9.2.4.1 Power Law Fluid Model

$$\sigma = k\dot{\gamma}^n \tag{1}$$

where k is the consistency index and n is the shear thinning index; $n < 1$.

By fitting the experimental data to equation 1, one can obtain k and n. The viscosity at a given shear rate can be calculated:

$$\eta = \frac{\sigma}{\dot{\gamma}} = \frac{k\dot{\gamma}^n}{\dot{\gamma}} = k\dot{\gamma}^{n-1} \tag{2}$$

9.2.4.2 Herschel–Bulkley General Model

Many systems show a dynamic yield value followed by a shear thinning behavior. The flow curve can be analyzed using the Herschel–Bulkley equation [16]:

$$\sigma = \sigma_\beta + k\dot{\gamma}^n \tag{3}$$

When $\sigma_\beta = 0$, equation 3 reduces to the power fluid model. The Herschel–Bulkley equation fits most flow curves with a good correlation coefficient, hence it is the most widely used model.

9.2.4.3 **Casson Model**

This is a semiempirical linear parameter model that has been applied to fit the flow curves of many paints and printing ink formulations [17]:

$$\sigma^{\frac{1}{2}} = \sigma_C{}^{\frac{1}{2}} + \eta_C{}^{\frac{1}{2}}\gamma^{\frac{1}{2}} \tag{4}$$

Thus a plot of $\sigma^{\frac{1}{2}}$ versus $\dot{\gamma}^{\frac{1}{2}}$ should give a straight line from which σ_C and η_C can be calculated from the intercept and slope of the line.

One should be careful in using the Casson equation since straight lines are only obtained from the results above a certain shear rate.

9.2.4.4 **Cross Equation**

This can be used to analyze the flow curves of shear thinning systems that show a limiting viscosity $\eta(0)$ in the low shear rate regime and another limiting viscosity $\eta(\infty)$ in the high shear rate regime [18]:

$$\frac{\eta - \eta(\infty)}{\eta(0) - \eta(\infty)} = \frac{1}{1 + k\gamma^n} \tag{5}$$

where k is the consistency index and n is the shear thinning index.

These two regimes are separated by a shear thinning behavior, as shown schematically in Figure 9.4.

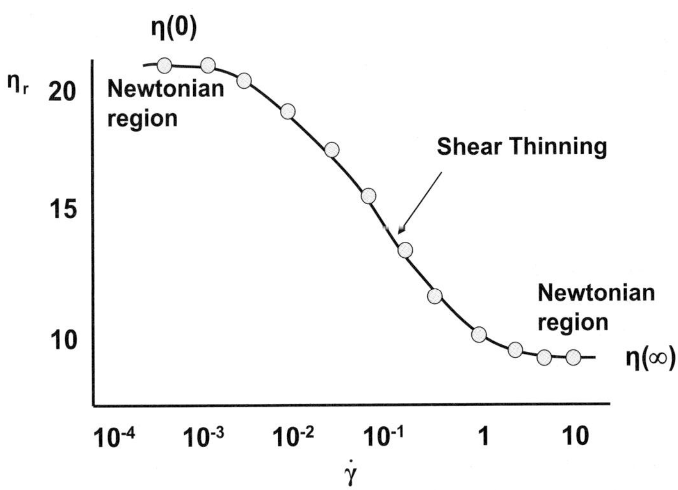

Figure 9.4 Viscosity versus shear rate for a shear thinning system.

9.2.5
Thixotropy

Most paints show thixotropic behavior in addition of being pseudoplastic. Thixotropy was first defined as an isothermal, reversible sol–gel–sol transition. This implies that the paint decreases in viscosity on shear (which is essential for ease of spreading and coating on the substrate), but builds up when the shearing process is stopped. Control of the time required for build-up of viscosity is essential for producing a paint film with the desirable properties. If the viscosity build-up is too fast, one may produce a film with the brush marks of the applicator. If the viscosity build-up is too slow, dripping and sagging may occur.

A schematic representation of the thixotropic behavior of a paint is shown schematically in Figure 9.5. If the flow could be measured without stirring (i.e. without breaking the structure), the curve AB could be produced [19]. However, on increasing the stress from 0 to 2W breakdown of the structure occurs, resulting in curve AC. When a continuous shear rate at 2W is applied over a period of time, the consistency of the paint decreases continuously from C to D, where it reaches a constant value, the lowest it can experience at the given shear stress 2W. Only a higher shear rate will be able to decrease the consistency further. If the shear is discontinued at point D, the build-up in the consistency of the paint to regain its original structure will follow along curve E, F or G, depending on the time which the particular paint requires for the building-up process.

Generally, two methods can be applied to study thixotropy in a paint formulation. The first and most commonly used procedure is the loop test, whereby the

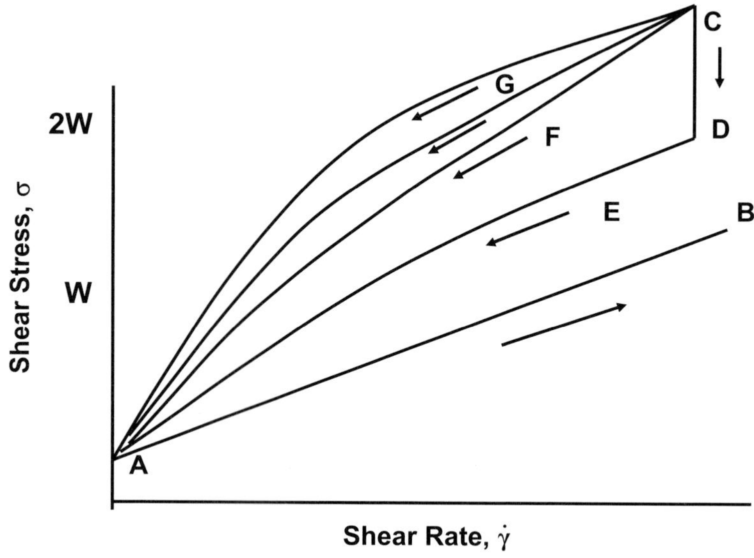

Figure 9.5 Schematic representation of thixotropic paint.

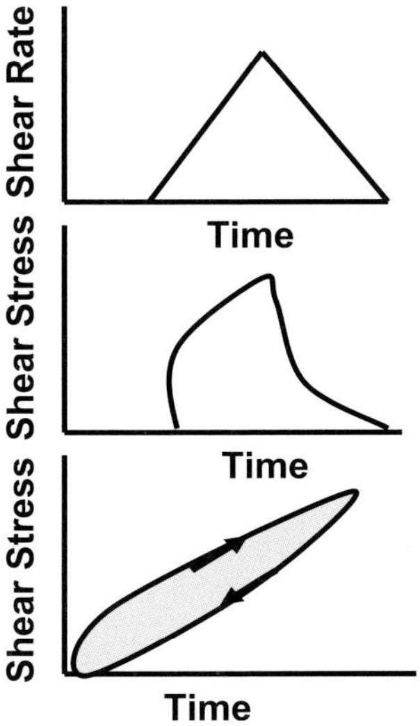

Figure 9.6 Loop test for studying thixotropy.

shear rate is increased continuously and linearly in time from zero to some maximum value and then decreased to zero in the same way [14]. This is illustrated in Figure 9.6.

The main problem with the above procedure is the difficulty of interpretation of the results. The non-linear approach used is not ideal for developing loops because by decoupling the relaxation process from the strain one does not allow the recovery of the material. However, the loop test reveals the qualitative behavior of the paint thixotropy.

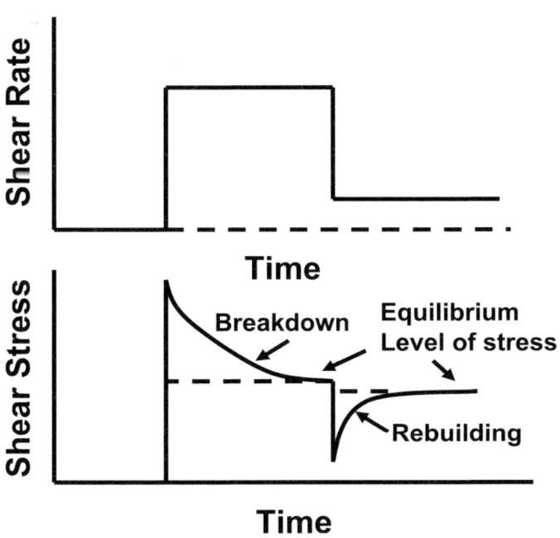

Figure 9.7 Step change test for studying thixotropy.

An alternative method for studying thixotropy is to apply a step change test, whereby the paint is suddenly subjected to a constant high shear rate and the stress is followed as a function of time, whereby the structure breaks down and an equilibrium value is reached [14]. The stress is further followed as a function of time to evaluate the rebuilding of the structure. A schematic representation of this procedure is shown in Figure 9.7.

Application of the above tests to a highly elastic paint is not straightforward, since there are contributions to the stress growth and decay from viscoelasticity. The occurrence of thixotropy implies that the flow must be taken into account when making predictions of flow behavior.

9.2.5.1 Transient Methods for Studying Paint Rheology

Two transient methods can be applied to study paint rheology: (i) stress relaxation after sudden application of strain and (ii) strain relaxation after sudden application of stress (creep measurements). Both of these techniques have been described in detail in Chapter 6 and here only a brief summary is given [12–15, 20].

Stress Relaxation Method In this case, a constant strain γ is applied within a very short period (that must be much smaller than the relaxation time of the sample) and the stress σ is followed immediately as a function of time. For a viscoelastic liquid (which is the case with many paint systems), the stress decreases exponentially with time t and reaches zero at infinite time. If the stress is divided by the applied constant strain, one obtains the stress relaxation modulus $G(t)$, which is related to the instantaneous modulus by the following expression:

$$G(t) = \frac{\sigma(t)}{\gamma} = \frac{\sigma(0)}{\gamma} \exp\left(-\frac{t}{\tau_m}\right) = G(0) \exp\left(-\frac{t}{\tau_m}\right) \qquad (6)$$

For a viscoelastic solid, the modulus reaches a limiting value G_e at long time (sometimes referred to as the equilibrium modulus). In this case, equation 6 has to be modified to account for G_e:

$$G(t) = G(0) \exp\left(-\frac{t}{\tau_m}\right) + G_e \qquad (7)$$

Note that according to equation 6, $t = \tau_m$ when $\sigma(t) = \sigma(0)/e$ or when $G(t) = G(0)/e$. This shows that stress relaxation can be used to obtain the relaxation time for a viscoelastic liquid.

Constant Stress (Creep) Method A constant stress σ is applied to the system (which may be placed in the gap between two concentric cylinders or a cone and plate geometry) and the strain (relative deformation) γ or compliance J ($= \gamma/\sigma$, Pa^{-1}) is followed as a function of time for a period t. At $t = t$, the stress is removed and the strain γ or compliance J is followed for another period t.

The above procedure is referred to as 'creep measurement'. From the variation of J with t when the stress is applied and the change in J with t when the stress is removed (in this case J changes sign), one can distinguish between viscous, elastic and viscoelastic response (see Chapter 6).

For viscoelastic response (as is the case with most paint systems), the following trend is observed: at $t = 0$, J shows a sudden increase and this is followed by a slower increase for the time applied. When the stress is removed, J changes sign and shows an exponential decrease with increase in time (creep recovery), but it does not reach zero as in the case of an elastic response.

9.2.6 Analysis of Creep Curves

9.2.6.1 Viscoelastic Liquid

The compliance $J(t)$ is given by two components an elastic component J_e that is given by the reciprocal of the instantaneous modulus and a viscous component J_v that is given by $t/\eta(0)$:

$$J(t) = \frac{1}{G(0)} + \frac{t}{\eta(0)} \tag{8}$$

The Maxwell relaxation time τ_M is given by

$$\tau_M = \frac{\eta(0)}{G(0)} \tag{9}$$

9.2.6.2 Viscoelastic Solid

In this case, complete recovery occurs and the system is characterized by a Kelvin retardation time τ_k that is also given by the ratio $\eta(0)/G(0)$.

9.2.6.3 Berger Model (Maxwell + Kelvin)

This model represents most practical paints, consisting of a Maxwell element and a Kelvin element. The modulus of the spring in the Maxwell element is G_1 and the viscosity in the dash-pot is η_1. The Maxwell relaxation time is η_1/G_1. The modulus of the spring in the Kelvin element is G_2 and the viscosity in the dash-pot is η_2. The Kelvin retardation time is η_2/G_2. The Berger model gives an instantaneous elastic response from G_1 and a continuous viscous response from η_1.

More complex models can be introduced, such as the generalized Maxwell model where several elements with different relaxation times are introduced. The generalized Kelvin model also consists of several Kelvin elements with different retardation times.

9.2.6.4 Creep Procedure

In creep experiments, one starts with a low applied stress (below the critical stress σ_{cr}; see below) at which the system behaves as a viscoelastic solid with complete

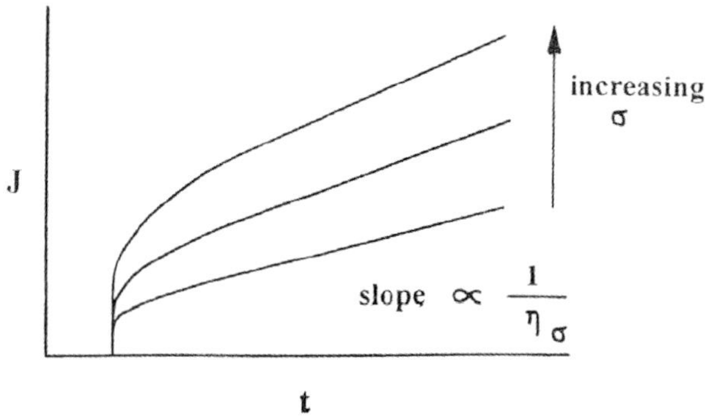

Figure 9.8 Creep curves at increasing applied stress. Creep measurements (constant stress) can be used to obtain the residual shear viscosity.

recovery. The stress is gradually increased and several creep curves are obtained. Above σ_{cr}, the system behaves as a viscoelastic liquid showing only partial recovery. Figure 9.8 shows a schematic representation of the variation of compliance J with time t at increasing σ (above σ_{cr}).

From the slopes of the lines, one can obtain the viscosity η_σ at each applied stress. A plot of η_σ versus σ is shown in Figure 9.9. This shows a limiting viscosity $\eta(0)$ below σ_{cr}, and above σ_{cr} the viscosity shows a sharp decrease with further increase in σ.

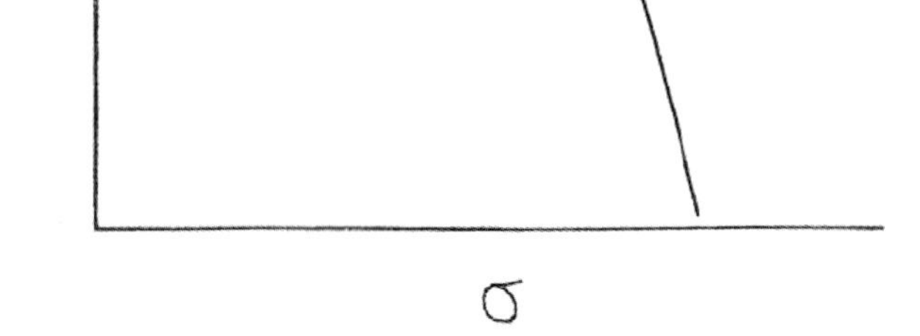

Figure 9.9 Variation of viscosity with applied stress. Critical stress is a useful parameter (related to yield stress) as it denotes the stress at which the structure 'breaks down'.

$\eta(0)$ is referred to as the residual or zero shear viscosity, which is an important parameter for predicting sedimentation. σ_{cr} is the critical stress above which the structure 'breaks down'. It is sometimes referred to as the 'true' yield stress.

9.2.7 Dynamic (Oscillatory) Techniques

This concerns the response of the material to an oscillating stress or strain. When a sample is constrained in, say, a cone and plate or concentric cylinder assembly, an oscillating strain at a given frequency ω (rad s^{-1}) ($\omega = 2\nu\pi$, where ν is the frequency in cycles per second or Hz) can be applied to the sample. After an initial start-up period, a stress develops in response of the applied strain, i.e. it oscillates with the same frequency. The change of the sine waves of the stress and strain with time can be analyzed to distinguish between elastic, viscous and viscoelastic response. Analysis of the resulting sine waves can be used to obtain the various viscoelastic parameters, as discussed below.

Three cases can be considered. (i) Elastic response, whereby the maximum of the stress amplitude is at the same position as the maximum of the strain amplitude (no energy dissipation). In this case, there no time shift between stress and strain sine waves. (ii) Viscous response, whereby the maximum of the stress is at the point of maximum shear rate (i.e. the inflection point) where there is maximum energy dissipation. In this case, the strain and stress sine waves are shifted by $\omega t = \pi/2$ (referred to as the phase angle shift δ, which in this case is 90°). (iii) Viscoelastic response, where the phase angle shift δ is greater than 0° but less than 90°.

9.2.7.1 Analysis of Oscillatory Response for a Viscoelastic System

Let us consider the case of a viscoelastic system. The frequency ω is in rad s^{-1} and the time shift between strain and stress sine waves is Δt. The phase angle shift δ is given by (in dimensionless units of radians)

$$\delta = \omega \Delta t \tag{10}$$

As discussed in Chapter 6, for a perfectly elastic solid $\delta = 0$, for a perfectly viscous liquid $\delta = 90°$ and for a viscoelastic system $0° < \delta < 90°$.

The ratio of the maximum stress σ_0 to the maximum strain γ_0 gives the complex modulus $|G^*|$:

$$|G^*| = \frac{\sigma_0}{\gamma_0} \tag{11}$$

$|G^*|$ can be resolved into two components: storage (elastic) modulus G', the real component of the complex modulus, and loss (viscous) modulus G'', the imaginary component of the complex modulus:

$$|G^*| = G' + iG'' \tag{12}$$

where i is the imaginary number, which is equal to $(-1)^{\frac{1}{2}}$.

The complex modulus can be resolved into G' and G'' using vector analysis and the phase angle shift δ as described in Chapter 6:

$$G' = |G^*| \cos \delta \tag{13}$$

$$G'' = |G^*| \sin \delta \tag{14}$$

$$\tan \delta = \frac{G''}{G'} \tag{15}$$

Dynamic viscosity η':

$$\eta' = \frac{G''}{\omega} \tag{16}$$

Note that $\eta \to \eta(0)$ as $\omega \to 0$.

Both G' and G'' can be expressed in terms of frequency ω and Maxwell relaxation time τ_m by

$$G'(\omega) = G\frac{(\omega\tau_m)^2}{1 + (\omega\tau_m)^2} \tag{17}$$

$$G''(\omega) = G\frac{\omega\tau_m}{1 + (\omega\tau_m)^2} \tag{18}$$

In oscillatory techniques, one has to carry two types of experiments: strain sweep, where the frequency ω is kept constant and G^*, G' and G'' are measured as a function of strain amplitude; frequency sweep, where the strain is kept constant (in the linear viscoelastic region) and G^*, G' and G'' are measured as a function of frequency.

9.2.7.2 **Strain Sweep**

The frequency is fixed at, say, 1 Hz (or 6.26 rad s^{-1}) and G^*, G' and G'' are measured as a function of strain amplitude γ_0. G^*, G' and G'' remain constant up to a critical strain γ_{cr}. This is the linear viscoelastic region where the moduli are independent of the applied strain. Above γ_{cr}, G^* and G' start to decrease whereas G'' starts to increase with further increase in γ_0. This is the non-linear region. γ_{cr} may be identified with the critical strain above which the structure starts to 'break down'. It can also be shown that above another critical strain, G'' becomes higher than G'. This is sometimes referred to as the 'melting strain' at which the system becomes more viscous than elastic.

9.2.7.3 Oscillatory Sweep

The strain γ_0 is fixed in the linear region (taking a mid-point, i.e. not a too low strain where the results may show some 'noise' and far from γ_{cr}). G^*, G' and G'' are then measured as a function of frequency (a range of 10^{-3}–10^2 rad s^{-1} may be chosen depending on the instrument and operator patience). One can identify a characteristic frequency ω^* at which $G' = G''$ (the 'cross-over point'), which can be used to obtain the Maxwell relaxation time τ_m:

$$\tau_m = \frac{1}{\omega^*} \tag{19}$$

In the low-frequency regime, i.e. $\omega < \omega^*$, $G'' > G'$. This corresponds to a long-time experiment (time is the reciprocal of frequency) and hence the system can dissipate energy as viscous flow. In the high-frequency regime, i.e. $\omega > \omega^*$, $G' > G''$. This corresponds to a short-time experiment where energy dissipation is reduced.

At sufficiently high frequency $G' \gg G''$, $G'' \to 0$ and $G' \approx G^*$. The high-frequency modulus $G'(\infty)$ is sometimes referred to as the 'rigidity modulus', where the response is mainly elastic.

For a viscoelastic solid, G' does not become zero at low frequency. G'' still shows a maximum at intermediate frequency.

9.2.7.4 Cohesive Energy Density, E_c

The cohesive energy density, which is an important parameter for identification of the 'strength' of the structure in a dispersion can be obtained from the change of G' with γ_0:

$$E_c = \int_0^{\gamma_{cr}} \sigma \, d\gamma \tag{20}$$

where σ is the stress in the sample, which is given by

$$\sigma = G'\gamma \tag{21}$$

$$E_c = \int_0^{\gamma_{cr}} G'\gamma_{cr} \, d\gamma = \frac{1}{2}\gamma_{cr}{}^2 G' \tag{22}$$

Note that E_c is given in J m^{-3}.

9.2.8 Normal Force

Normal stresses may be responsible for the flow behavior of some paint formulations. The best known and certainly the most dramatic effect is the rod-climbing phenomenon, referred to as the 'Weissenberg effect'. It is observed when a rotating rod is dipped into a squat vessel containing an elastic liquid. Whereas a Newtonian liquid would be forced to the rim of the vessel by inertia, and thus produce a free surface that is higher at the rim than near the rod, the elastic liquid produces a free surface that is much higher near the rod. The Weissenberg effect may be viewed as a direct consequence of normal stress which acts like a hoop stress around the rod. This stress causes the liquid to 'strangle' the rod and hence move along it.

Consider, for example, a material possessing both elasticity and viscosity that is placed in the gap between parallel plates. If the upper plate is moved at a constant velocity v (or shear rate $\dot{\gamma}$), then due to the presence of elasticity (resistance to continuous deformation exerted by the material), the total force exerted on the moving plate is at an angle to the direction of motion. This total force can be resolved into its components, which include a force parallel to and in the same plane as the plate, and a component in a plane vertical to the plane of motion and at right-angles to the direction of motion. The latter is the normal force. In practical terms, the normal force tries to push the plates apart while there is motion. In practical instruments, the moving plate must either be held rigidly in the vertical plane, or it can be allowed to move and kept in position by applying an equal and opposite restoring force to counteract the normal force. This approach allows one to measure the normal force.

9.2.9
Extensional (Elongational) Viscosity

The importance of measurement of viscosity under extensional conditions is well known in the area of fiber formation, i.e. strongly spinnable materials. However, it has been recognized that measurement of extensional viscosity is of direct relevance in many other areas such as ink-jet printers, roll mills, blade coating, curtain coating, emulsions and suspensions. Thus, measurement of the extensional viscosity of paint systems could be important for several applications. With many polymer solutions, the extensional viscosity can be several orders of magnitude higher than the shear viscosity. The same may apply to some paint systems which contain high molecular weight polymers (rheology modifiers). Unfortunately, measurement of extensional viscosity is not easy, although recently some manufacturers have designed special instruments for such experiments.

9.3
Application of Rheological Techniques to Paint Formulations

Any change in the physical or chemical characteristics of a paint formulation is directly reflected in its flow characteristics or rheology. These changes can occur as a result of aging, temperature changes, application of shear, type of dispersion, extent of grinding and mixing and addition of special surfactants for some applications. Below, a summary of the possible physical changes that may occur and their evaluation using rheological techniques is presented.

9.3.1
Aging

The aging of a paint (during storage) can cause an increase or decrease in consistency. This change in consistency could be due to flocculation that may result from desorption of the dispersant during storage or simply to temperature fluctuations. Some chemical changes may also occur as a result of the reaction between the solid and liquid phases [19].

The change in consistency can be followed by measuring the flow curves at various intervals of time, as illustrated in Figure 9.10. This presentation illustrates that the consistency of the paint increases on storage, which is most likely due to flocculation. The flow curves can be analyzed using the models described in Chapter 6. Using these models, one can calculate the yield value and the viscosity as a function of storage time. Both parameters show an increase with increase in storage time.

Another important investigation is to follow the thixotropic behavior of the paint during storage. As mentioned earlier, thixotropy can be investigated using the thixotropic loop or the step change method. Any flocculation will also be accompanied by a change in the thixotropic behavior.

Clearly, a physically stable paint should show no change in its rheological behavior on storage for at least 6 months and also at various temperatures to which the paint system will be subjected.

A more sensitive method for following the change in consistency on storage is constant stress or creep measurements. As discussed earlier, these measurements

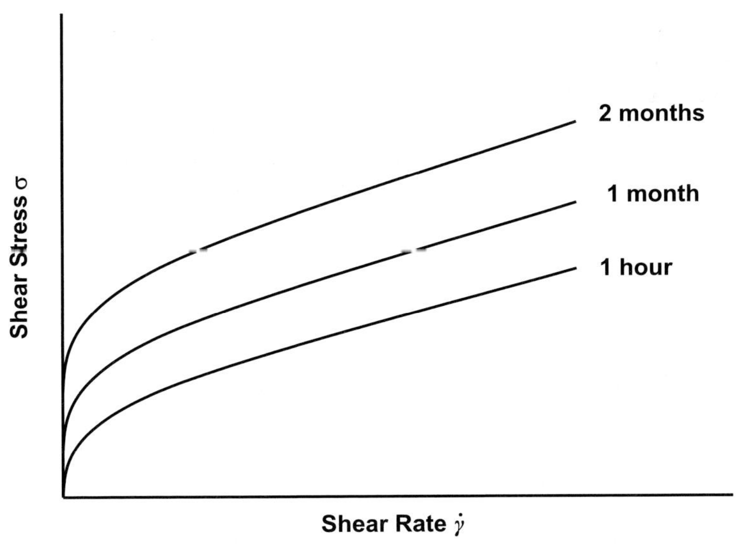

Figure 9.10 Schematic representation of the change in the flow curve of a paint system at various times.

allow one to obtain the residual (or zero shear) viscosity $\eta(0)$ and also the critical stress σ_{cr} above which the structure breaks down. Any increase in $\eta(0)$ and σ_{cr} indicates flocculation of the paint on storage.

A third and sensitive method for following the change in consistency on storage is the dynamic (oscillatory) method described above. By following the change in elastic modulus and cohesive energy during storage, one can obtain information on the flocculation of the paint.

9.3.2
Temperature

The consistency or the viscosity η of most paint formulations decreases with increase in temperature. However, in some cases the viscosity may show an increase with increase in temperature when the latter reaches a critical value. In most cases this is due to sudden flocculation of the paint above a critical temperature (referred to as the critical flocculation temperature, CFT). This flocculation may result from a decrease in solvency of the chains to worse than a θ-solvent above a critical temperature. Alternatively, the flocculation may occur as a result of desorption of the dispersant at high temperature due to the sudden increase in its solubility.

Due to the above changes in the state of the paint with change in temperature, the viscosity–temperature relationship seldom follows an Arrhenius plot (which shows a linear relationship between log η and $1/T$, where T is the absolute temperature.

A rapid technique to study the effect of temperature changes on the flocculation of a paint formulation is to carry out temperature sweep experiments, running the samples from say 5 to 50 °C. The trend in the variation of σ_β and η_{pl} with temperature can quickly give an indication on the temperature range in which a paint remains stable (within that temperature range σ_β and η_{pl} remain constant).

9.3.3
Dispersion and Ingredients

Most paint formulations consist of a mixture of suspended particles (pigments) and emulsion droplets (latex particles that are liquid-like at room temperature), referred to as suspoemulsions. The continuous medium with viscosity η_0 may be simply an aqueous phase in which several ingredients are dissolved or could be a non-aqueous (oil) phase that may consist of two or more miscible oils. For non-Newtonian systems (which is the case with paint systems), some empirical equations can be established to relate the plastic viscosity η_{pl} and yield value σ_β to the volume fraction of the disperse phase ϕ [20]:

$$\eta_{pl} = (\eta_0 + A)\exp(B\phi) \tag{23}$$

$$\sigma_\beta = M\exp(n\phi) \tag{24}$$

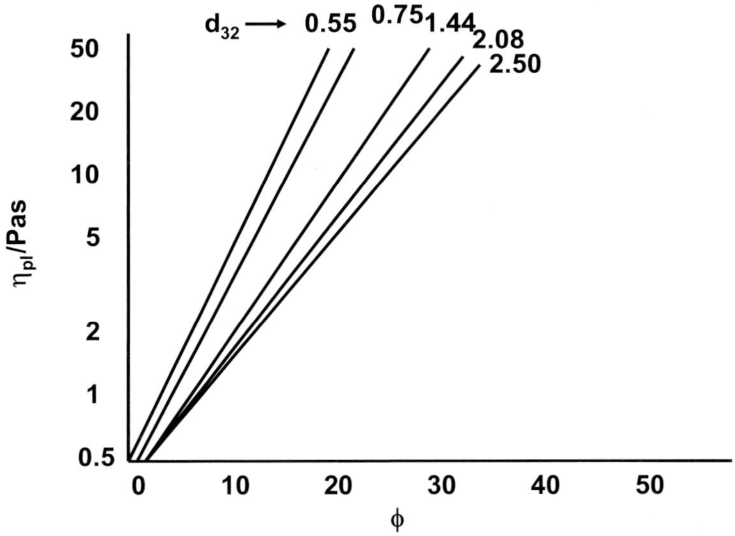

Figure 9.11 Variation of log η_{pl} with ϕ for leaded ZnO suspensions with different particle diameters d_{32}.

where B, M and n are related to particle size, shape and surface. A is independent of particle size but may depend on particle shape and surface; it may be related to particle–particle interaction in the dispersion.

Equations 23 and 24 predict a linear relationship between log η_{pl} or log σ_β and ϕ. This is illustrated in Figures 9.11 and 9.12, which also show the effect of aver-

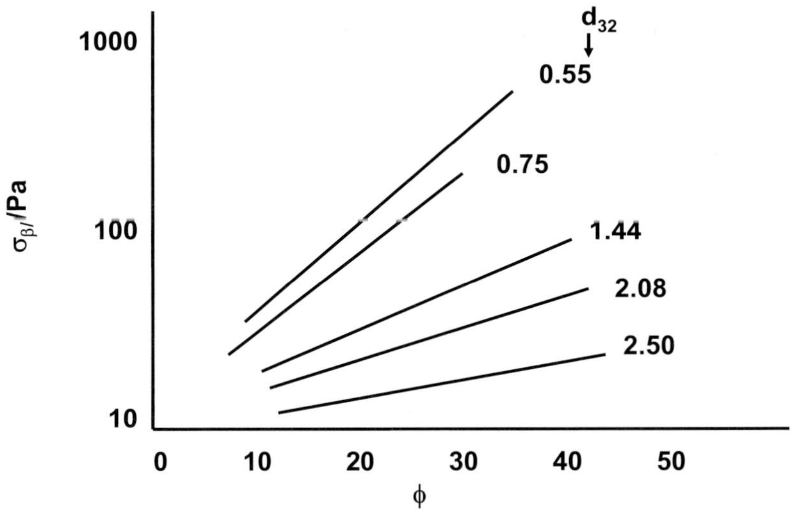

Figure 9.12 Variation of yield value σ_β with volume fraction ϕ for ZnO suspensions with different particle diameters d_{32}.

age particle diameter (volume to surface ratio d_{32}) of the pigment [20]. The smaller the size, the higher is the slope. This is illustrated in Figure 9.13, which shows the variation of B and n with particle diameter. It is clear that both B and n increase with decrease in d_{32}.

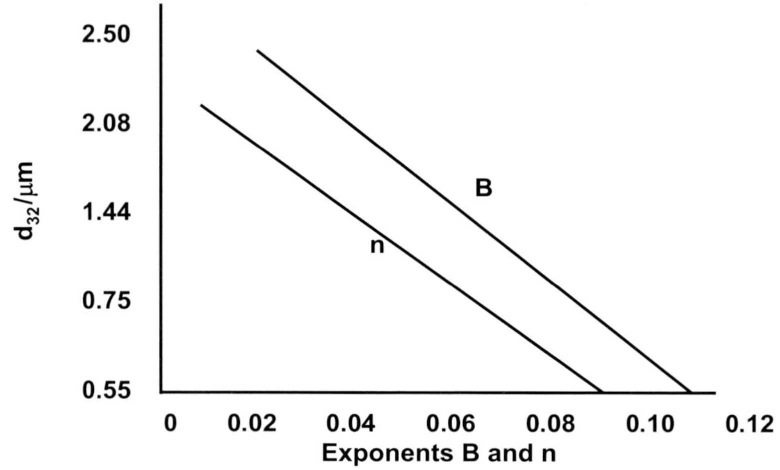

Figure 9.13 Variation of B and n with average particle diameter d_{32}.

9.3.4 Effect of Surface-active Agents and Dispersants

As mentioned in Chapter 3, surface-active agents are added to aid the wetting of the pigment powder. Dispersing agents (mostly polymeric surfactants) are also added to stabilize the particles against aggregation. Both materials affect the viscosity and yield value of the final paint dispersion by adsorption at the solid/liquid interface. The main purpose of wetters and dispersants is to produce 'better dispersion' by causing disaggregation and deflocculation. Disaggregation is a mechanical or chemical separation of single particles in an aggregate. The aggregate consisting of these unit particles is 'glued' together, thus preventing the liquid from penetrating into the aggregated mass and thus surrounding each unit particle. Deflocculation, on the other hand, can only be effected by the use of a dispersing agent. The mechanical force does not change the state of flocculation. A flocculate is a 'loose' but connected structure of particles, where the particles are far enough apart to permit the liquid to surround them. However, the particles are sufficiently close to each other with strong van der Waals attraction. Hence the dispersion will not flow until enough shear stress is applied to overcome these attractive forces. This shear stress is proportional to the yield value. A dispersant that is strongly adsorbed to the particle surface and providing sufficient repulsive forces can overcome the van der Waals attraction, thus causing a marked reduction in the yield value.

It should be mentioned that controlled flocculation of a pigment dispersion can be desirable to prevent settling and formation of hard sediments and to control the surface finish of a coating. For pigments dispersed in oil, small quantities of a polar liquid such as alcohol, glycerol or butanol are used as a flocculating agent. For hydrophilic pigments suspended in aqueous media, oils and oil-soluble agents, such as lecithin, can induce flocculation.

9.3.5
Grinding and Mixing

In general, an increase in the viscosity of a dispersion results in an increase in the efficiency of milling. This is represented schematically for a three-roller mill in Figure 9.14, which shows the variation of milling time with the plastic viscosity of the dispersions measured using a rotational viscometer [20]. It should be mentioned that the trend shown in Figure 9.14 gives only an indication since the operational viscosity in the three-roller mill is not the same as the plastic viscosity measured using a rotational viscometer. Apparently, the yield value does not affect the grinding efficiency, provided that it is low enough that the material flows readily from the feed rollers.

In ball milling, the viscosity of the dispersion also plays an important role. A practical viscosity for good operation depends on the nature of the balls. When using steel ball mills, a high viscosity (up to 20 Pa s) can be used. With pebble and porcelain ball mills, a lower viscosity is required since the weight of the grinding medium is lower. It should also be mentioned that the viscosity measured before mixing is substantially different from that existing during the mixing operation. The flow properties of the dispersion during the process of grinding change as a result of increase in temperature, increase in wetting, increased degree of aggregation and improved interaction between the solid and liquid phases.

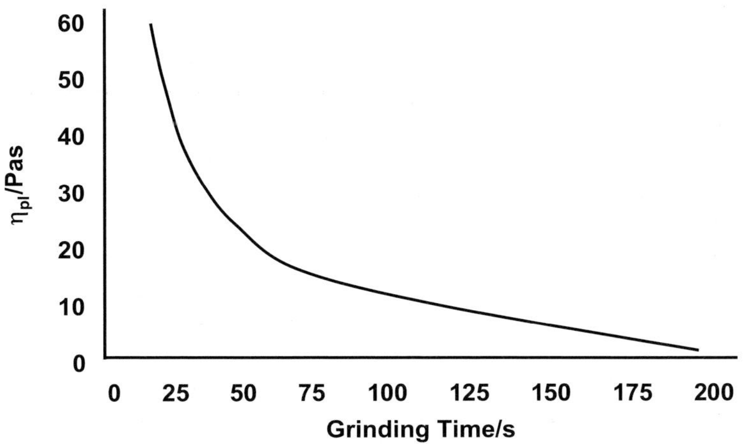

Figure 9.14 Schematic representation of the effect of dispersion viscosity on grinding time.

Microscopic investigations showed an increase in disaggregation during milling and this is accompanied by increase in color strength [21]. This is illustrated schematically in Figure 9.15, which shows the change in plastic viscosity, yield value and color strength for a carbon black dispersion in mineral oil. Because the shear rate is much higher during milling than the maximum value measured in a rotational viscometer (usually of the order of 1000 s^{-1}), the viscosity of a pseudoplastic plastic material will decrease substantially, whereas its yield value may increase during milling. The viscosity of a thixotropic dispersion will decrease substantially, whereas its yield value may increase during milling. Therefore, to evaluate the grinding performance, the consistency of the dispersion under the operational grinding conditions and at different steps of processing has to be determined. This may require measurement of the viscosity at much higher shear rates than encountered with rotational viscometers, as determined, for example, using capillary viscometers (which can operate at much higher shear rates).

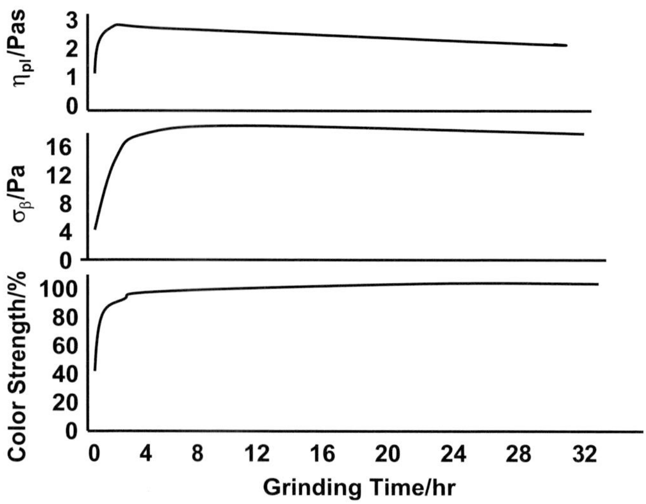

Figure 9.15 Change in flow properties and color strength of carbon black suspensions in mineral oil as a function of ball milling time.

9.4
Application of Rheology for Paint Evaluation

Rheology is perhaps one of the most powerful techniques for paint evaluation both during its formulation and in its manufacture. Before a paint is manufactured, its application is known and it is essential to control its flow properties for optimum operation and application. To control the flow properties of the paint, a rheometer must be selected to make flow measurements, which permits good interpretation of the flow properties of the paint. This allows the manufacturer to decide whether two batches of the same material or of different materials will have equal flow behavior under all conditions of operational application. The manufacturer could also predict from these flow measurements whether there is any difference in physical properties on paint application. In the preoperational stage, physical effects which occur in manufacturing and storing, such as temperature effects, evaporation, mixing procedures and shelf-life, must be studied. These physical effects can be correlated with the flow characteristics of the paint formulation. This allows one to achieve a more efficient and better controlled operation in manufacture and application.

At present, many paint manufacturers use one-point measurement for measurement of the consistency. This can be carried out, for example, using a simple Brookfield viscometer using one spindle at a given rotational speed. This one-point measurement can be misleading [20]. To illustrate this point, let us consider three systems, namely Newtonian, Bingham plastic and pseudoplastic (thixotropic), as illustrated in Figure 9.16. At a specific shear rate, all three systems show the same apparent viscosity, although their flow behavior (using the full shear–stress curves) are significantly different [20]. These systems will show entirely different behavior on application at the shear rate at which the apparent viscosity is the same. A study of the flow curves indicates that the Newtonian system will flow at extremely low shear rates, whereas the plastic and thixotropic systems will show a reluctance to do so because of their yield values. This is clearly reflected in the final film properties. Once the yield value is overcome, the viscosity of the paint may become even lower than that of a Newtonian system. At the operational high shear rates all the three systems show different viscosities.

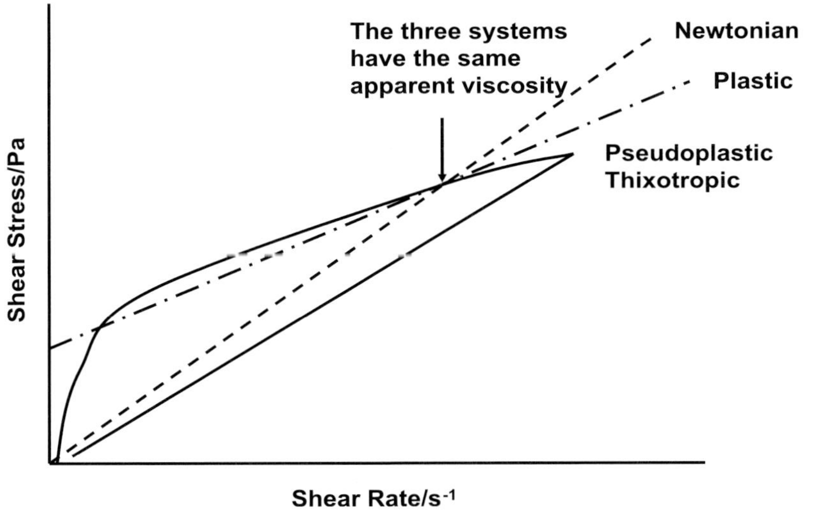

Figure 9.16 Schematic representation of the flow behavior of Newtonian, plastic and pseudoplastic (thixotropic) systems.

9.4.1 Flow in Pipes

Since most paints are pumped through pipes during manufacture and in application, it is essential to analyze their flow behavior in pipes. Two types of flow behavior must be considered: (i) laminar flow, where distinct layers of material would pass each other, and (ii) turbulent flow, where no distinct layers are observed and all layers mix with one another forming eddy currents, swirls and vortices. Whether the flow is laminar or turbulent depends on the dimensionless Reynolds number, R_e, which is given by the following equation:

$$R_e = \frac{vl\rho}{\eta} \tag{25}$$

where v is the mean velocity, l is an instrument length parameter, ρ is the density and η is the viscosity. For laminar flow $R_e \leq 2000$, whereas for turbulent flow $R_e \geq 2000$. Therefore, when the shear rate exceeds a critical value, the laminar flow changes to turbulent flow. This is reflected in the flow curve, which shows an apparent increase in shear stress above a critical shear rate. This does not mean that the viscosity of the material increases with increase in shear rate (as is the case with dilatant systems), but it indicates that with increasing shear rate the degree of turbulence increases, since part of the increased stress is used to increase the number of eddy currents rather than to increase the flow of the bulk material.

Turbulent flow in pipes, where l is replaced by the mean pipe diameter D, can occur when the Reynolds number exceeds 2000. Empirical equations have been established for turbulent flow of Newtonian materials [22, 23]. For smooth pipes:

$$\frac{1}{\sqrt{f}} = 2\log\left(\frac{R_e}{\sqrt{f}}\right) - 0.8 \tag{26}$$

and for rough pipes:

$$\frac{1}{\sqrt{f}} = 2\log\left(\frac{D}{2k}\right) \tag{27}$$

where f is the friction factor, given by

$$f = \frac{2DP}{\rho L v^2} = \frac{64}{R_e} \tag{28}$$

k is the grain diameter that indicates surface roughness, P is the pressure in the pipe and L is the overall length of the pipe. Hence for turbulent flow in smooth pipes the flow velocity depends on the Reynolds number and hence on the viscosity.

The pressure that is necessary to pump a material through a pipeline system at a given flow rate depends on the pressure loss in the total pipe line system. Pressure losses are incurred by the viscous resistance of the material in the straight pipeline and in the pipeline transitions such as bends, valves, elbows, pipe expansions and contractions. The viscous losses in the straight pipelines are frequently large compared with the pipe transitions, so that the latter can sometimes be neglected.

The pressure loss for an entire pipeline system is given by [20]

$$\Delta P = \rho \frac{v^2}{2} \left(\frac{L}{D} f + C_{\mathrm{L}} \right) \tag{29}$$

where C_{L} is the sum of all pressure loss coefficients obtained from all pipeline transitions in the pipeline system.

The flow of Newtonian materials in pipelines under laminar flow is well understood and is given by the Poiseuille equation:

$$\eta = \frac{\pi R^4 P}{8QL} \tag{30}$$

where Q is the volumetric flow rate ($\mathrm{m}^3\ \mathrm{s}^{-1}$) and R is the pipe radius.

In turbulent flow, the flow for Newtonian systems is given by equations 26 and 27.

The flow of non-Newtonian materials (as is the case with paints) in pipelines is not as well understood. However, Buckingham [24] derived the following equation for evaluating the plastic viscosity of a Bingham plastic system from the flow curve in a capillary viscometer (assuming end effects, kinetic energy effects and slippage flow are absent):

$$\eta_{\mathrm{pl}} = \frac{\pi P R^4}{8QL} \left[1 - \frac{8L\sigma_\beta}{3RP} + \frac{1}{3} \left(\frac{2L\sigma_\beta}{RP} \right)^4 \right] \tag{31}$$

Equation 31 can be used to determine the laminar flow of plastic materials in pipelines with R being the pipe radius. For pseudoplastic and dilatant materials, the power law equation can be used:

$$\eta = k\dot{\gamma}^{n-1} \tag{32}$$

where k is the consistency index and n is the shear thinning index ($n < 1$ for pseudoplastic materials).

For a Bingham plastic in laminar flow, the friction factor f is given by

$$f = \frac{64}{R_{\mathrm{e}}} \frac{Pl}{8s} \tag{33}$$

where *Pl* is the plasticity number, given by

$$Pl = \frac{\sigma_\beta D}{Uv} \tag{34}$$

where U is the coefficient of plastic viscosity and v is the velocity, and s is the ratio of the yield value to the shear stress at the wall. Since s is a function of *Pl*, the friction factor for Bingham plastics is fully determined from R_e and *Pl*.

For pseudoplastic materials in laminar flow, the friction factor is given by [20]

$$f = \frac{64}{R_e}\left(\frac{3+N}{4}\right) \tag{35}$$

where $N = 1/n$. Hence the friction factor for pseudoplastic materials is fully determined from R_e and N.

The shear rate in the pipeline for the flow of pseudoplastic materials is given by [20]

$$\dot{\gamma} = \frac{2v(N+3)}{D} \tag{36}$$

The apparent viscosity that is to be used in the Reynolds number has to be measured in the viscometer at the pipeline shear rate. This can be obtained by fitting the flow curve to the power law relationship given by equation 32.

In turbulent non-Newtonian flow, the friction factor is a unique function of the Reynold's number. For Bingham plastic systems, the Reynold's number is calculated by using the plastic viscosity since it remains constant with increasing shear rate. For pseudoplastic flow, the Reynold's number is calculated by using an estimated apparent viscosity that is obtained by extrapolation to infinite shear rate.

9.5
Examples of the Flow Properties of Some Commercial Paints

Most commercial paints have flow characteristics similar to those of thixotropic plastic materials [20]. Typical examples of the flow characteristics of some paints are given in Table 9.1.

The results in Table 9.1 were obtained at the same shear rate (using a Stormer-type concentric cylinder) and they clearly show similar flow characteristics with viscosities of less than 0.4 Pa s and yield values not exceeding 12 Pa.

The rheological behavior of a paint and a lacquer during and after application determines the smoothness and perfection of the resulting film surface. Paints are applied by brushing, dipping, flow coating and spraying. In all cases the flow-out of the material between the time of application and drying determines

Table 9.1 Flow properties of some commercial paints.

Product	η/Pa s	σ_β/Pa	Degree of thixotropy
Enamels, gloss	0.14–0.39	0–3	Nil to slight
Enamels, semi-gloss	0.10–0.35	5–12	Slight
Flat or matte paints	0.06–0.10	2–10	Slight to marked
Wall Water-dispersible	0.02–0.14	1–10	Slight to marked
Primers, metal	0.03–0.12	0–10	Nil to marked
Varnishes	0.09–0.29	0	Nil

the characteristics of the finished surface. The time required for flow-out of oils of different viscosities on a non-porous surface depends on the layer thickness as is illustrated in Table 9.2.

The results in Table 9.2 show that leveling to a smoother film takes less time with lower viscosity paints and for films with greater thickness. However, one should remember that the film viscosity continues to increase while drying and leveling proceed. Generally, the flow-out is usually complete in about 1 min if the viscosity does not exceed 0.1 Pa s. Most paints require longer flow-out times since the viscosity is higher than 0.1 Pa s at 1 min after deposition.

The shear rate produced by brushing paints on a surface was estimated to range from 130 to 260 s^{-1}. This value can be obtained by determining the shear rate at which the order of apparent viscosities of thyrotrophic paints exhibiting entirely different flow behavior coincide with the order of ease of flow indicated by practical brushing. Alternatively, the shear rate can be calculated by assuming that the brushing velocity is 0.2 m s^{-1} for a distance between the substrate and the brush of about 0.001 m (this gives a shear rate of 200 s^{-1}). However, other workers assumed much higher shear rates (about 100 times higher). Various other investigators tried to correlate the flow behavior of paints with their brushing behavior. In general, paints have good brushing properties if the plastic

Table 9.2 Approximate time of leveling of oil films on a non-porous substrate as a function of viscosity and film thickness.

η/Pa s	Leveling time/min	Film thickness/mm
<0.05	<1	0.5–5.0
0.20	5	0.5
0.20	<0.5	$\gg 0.5$
1.0	>120	0.5
1.0	10	$\gg 0.5$

viscosity ranges from 0.2 to 0.5 Pa s and the yield value from 40 to 140 Pa. Both the plastic viscosity and yield value increase with increase in solvent evaporation (particularly for non-aqueous paints based on volatile solvents). This is illustrated in Table 9.3 for brushed black enamel paint [25].

A paint is considered to have good brushing properties when all brush marks disappear during the drying process. This might be achieved by rapid flow-out caused by low viscosity and low yield value or by taking advantage of the thixotropic behavior of the paint. Low viscosity and low yield value often lead to sag marks and 'curtains', since the low consistency causes the paint to continue to flow after application until drying is sufficiently advanced. Sagging can also appear if the film is applied in a layer that is too thick. The thixotropic properties have the advantage of giving the paint the low operational consistency needed for initial flow-out in order to prevent the brush marks and to obtain good leveling. At the same time, they provide the paint with a mechanism for increasing the consistency by means of the thixotropic build-up of the structure that can be effective immediately after application. However, a too rapid thixotropic build-up can be detrimental because leveling and flow-out are not instantaneous. It has been shown [26, 27] that thixotropic paints which decrease rapidly in consistency with increase in shear rate but rebuild the structure slowly exhibit the best leveling characteristics. A too rapid increase in thixotropic structure can produce poor leveling.

The viscosity of paints and lacquers that are applied by spraying (using, for example, a spray-gun) range from 0.04 to 0.12 Pa s. For hot sprays that are applied at about 75 °C, the viscosity should be adjusted to about 0.2 Pa s at room temperature, so that it will be less than 0.1 Pa s at the spraying temperature. The viscosity of hot plastic paints that are sprayed at temperatures around 150 °C should be less than 0.15 Pa s. The viscosity of the paint is usually adjusted by addition of solvent. The viscosity of the bulk paint has to be very low for spray operations, since it will increase substantially on evaporation of the solvent during the spraying process (from the time it leaves the spray nozzle to the time when it hits the substrate on which it is to be coated). The increase in viscosity depends on the volatility of the solvent.

When applying a paint by spraying, the droplet size of the spray will greatly influence the appearance of the finished surface. A fine spray will produce a glossy

Table 9.3 Change in flow properties of a brushed black enamel.

Time/min	η_{pl}/Pa s	σ_β/Pa	Type of flow
0	0.19	0	Newtonian
2.5	1.4	2.4	Plastic
5.0	3.0	52.0	Thyrotrophic plastic
7.5	16.7	620.0	Thixotropic plastic

surface, whereas a coarse spray will gave a matte finish, unless leveling occurs before drying is completed. The drop size of the spray depends on the physical properties of the paint, such as its surface tension, viscosity and density. Other parameters such as the flow rate, the relative velocities of the liquid and air also affect the spray droplet spectrum.

References

1 T.A. Strivens, in *Paint and Surface Coatings*, ed. R. Lambourne, Ellis Horwood, Chichester (1987).
2 J.E. Glass, *Coat. Technol.*, *50*, 56 (1978).
3 T.C. Patton, *Paint Flow and Pigment Dispersion*, Wiley-Interscience, New York (1979).
4 Y. Kuge, *Coat. Technol.*, *55*, 59 (1983).
5 N.D.P. Smith, S.E. Orchard and A. Rhind-Tutt, *J. Oil Colour Chem. Assoc.*, *44*, 618 (1961).
6 J.A.R. Pearson, *J. Fluid Mech.*, *7*, 481 (1960).
7 M.D. Savage, *J. Fluid Mech.*, *80*, 473 (1977).
8 J.E. Glass, *J. Oil Colour Chem. Assoc.*, *58*, 169 (1975).
9 J.S. Dodge, *J. Paint Technol.*, *44*, 72 (1972).
10 C.I. Snow, *Official Digest*, (392) 907 (1957).
11 R.R. Myers, *J. Polym. Sci.*, *Part C*, *35*, 3 (1971).
12 J.R. van Wazer, J.W. Lyons, K.Y. Kim and R.E. Cowell, *Viscosity and Flow Measurements*, Interscience, New York (1983).
13 R.W. Whorlow, *Rheological Techniques*, John Wiley and Sons, Inc., New York (1980).
14 H.A. Barnes, J.F. Hutton and K. Walters, *An Introduction to Rheology*, Elsevier, Amsterdam (1989).
15 J.W. Goodwin and R.W. Hughes, *Rheology for Chemists*, Royal Society of Chemistry, Cambridge (2000).
16 W.H. Herschel and R. Bulkley, *Proc. Am. Soc. Test. Mater.*, *26*, 621 (1926); *Kolloid Z.*, *39*, 291 (1926).
17 N. Casson, in *Rheology of Disperse Systems*, ed. C.C. Mill, Pergamon Press, New York, pp. 84–104 (1959).
18 M.M. Cross, *J. Colloid Interface Sci.*, *20*, 417 (1965).
19 R.N. Weltman, in *Rheology*, Vol. 3, ed. F.R. Eirich, Academic Press, New York, Chapter 6 (1960).
20 J.D. Ferry, *Viscoelastic Properties of Polymers*, John Wiley & Sons, Inc., New York (1980).
21 E.K. Fischer, *Ind. Eng. Chem.*, *33*, 1465 (1941).
22 H. Rouse and J.W. Howe, *Basic Mechanics of Fluids*, John Wiley & Sons, Inc., New York (1953).
23 E.N.da C. Andrade, *Viscosity and Plasticity*, Chemica Publishing, New York (1952).
24 E. Buckingham, *Proc. Am. Soc. Test. Mater.*, *21*, 1154 (1921).
25 E.K. Fischer, *J. Colloid Sci.*, *5*, 271 (1950).
26 B. Saunders, *J. Oil Colour Chem. Assoc.*, *31*, 95 (1948).
27 M.E.D. Jarret, *J. Oil Colour Chem. Assoc.*, *31*, 337 (1948).

Index

Colloids and Interface Science Series, Vol. 6
Colloids in Paints. Tharwat F. Tadros

ISBN: 978-3-527-31466-9

f